Copyright © 2015 by Accepted, Inc.

ALL RIGHTS RESERVED. By purchase of this book, you have been licensed one copy for personal use only. No part of this work may be reproduced, redistributed, or used in any form or by any means without prior written permission of the publisher and copyright owner.

Questions regarding permissions or copyrighting may be sent to support@acceptedinc.com.

Table of Contents

Introduction

Overview — 7

Newtonian Mechanics

Units and Fundamental Concepts — 11

Vector Math — 15

Newton's Laws — 25

Kinematics and Projectile Motion — 31

Friction — 45

Work, Energy and Power — 51

Momentum — 59

Circular Motion — 67

Springs and Oscillations — 75

Fluid Mechanics and Thermal Physics

Fluid Mechanics — 83

Heat Transfer in Physics — 93

Electricity and Magnetism

Electric Point Charges — 101

Direct Current (DC) Circuit Analysis — 111

Magnetic Fields — 115

Waves and Optics

Traveling Waves and Sound - - - - - - - 121

Light Refraction and Optics - - - - - - - 129

Reflection and Refraction - - - - - - - 133

Lenses - - - - - - - - - 139

Atomic and Nuclear Physics

Nuclear Physics and the Model of the Atom - - - - - 143

Practice Examination

Free Response Questions: A Primer - - - - - - 149

Quiz 1 - - - - - - - - - 155

Answer Key 1 - - - - - - - - 187

Quiz 2 - - - - - - - - - 217

Quiz 3 - - - - - - - - - 277

Image Credits - - - - - - - - 335

Introduction

Exam Overview

Welcome to the study guide! The Advanced Placement (AP) Physics Examination is designed to challenge a student's knowledge of the content of a college-level physics course.

This guide is not designed to replace an AP Physics course, but it will help you to review and prepare for the examination. If you have not taken the course, you can use this guide as a general aid, supplemented with other materials like a college-level textbook. There is no specific or required curriculum for the course; however, there is a range of material typically included in a basic introductory college course on chemistry that will be on the examination.

Test Content

The test consists of multiple-choice questions (MCQs), as well as free response questions, in five different content areas:

1. Newtonian Mechanics (35%)
2. Fluid Mechanics and Thermal Physics (15%)
3. Electricity and Magnetism (25%)
4. Waves and Optics (15%)
5. Atomic and Nuclear Physics (10%)

Scoring

The test is scored on a scale of 1 to 5. A score of 5 means you are extremely well qualified to receive college credit, while a score of one means you are not qualified to receive college credit. While colleges and universities use scores differently, a score of 4-5 is equivalent to an A or B. A score of 3 is approximately similar to a C, while a score of 1-2 is comparable to a D or F. The examination is scored on a curve, and is adjusted for difficulty annually. This means that your test score (1-5) is equivalent to the same scores from tests in different years. The curve is different each year, depending upon the test.

Scores of 4 to 5 are widely accepted by colleges and universities; however, scores of 3 or lower may provide less credit or none at all. Elite schools may require a score of

5 for credit, and some schools vary the required score depending upon the department. You will need to review the AP policies at your college or university to better understand scoring requirements and credit offered.

Studying

This guide includes practice tests. While these tests familiarize you with the format and type of questions you will see on the test, they can also help you learn to employ your timing and test strategies and reduce your anxiety on test day.

Take the practice tests in a quiet, comfortable environment. Mimic the test environment as much as possible. Time the sections correctly and avoid getting up and down, fetching a drink, or having a snack. Plan to take the entire practice test in a single setting, just as you would on test day. Even if you can't do this for each of the practice tests, make time to do it for at least one of them, so you've had the full test experience once before the big day.

Staying Calm, Cool and Collected

Conquering test anxiety can help you succeed. Test anxiety is common and, if it's mild, can help keep you alert and on-task. If you're like most folks who get a little bit of anxiety, here are some tips to help you calm your nerves:

- Allow plenty of time for test preparation. Work slowly and methodically. Cramming doesn't help and will leave you depleted and exhausted.
- Remember to stay healthy. Sleep enough, eat right, and get regular exercise.
- Practice breathing exercises to use on test day to help with anxiety. Deep breathing is one of the easiest, fastest and most effective ways to reduce physical symptoms of anxiety.

While these strategies won't eliminate test anxiety, they can help you to reach exam day at your mental best, prepared to succeed.

The night before the test, just put away the books. More preparation isn't going to make a difference. Read something light, watch a favorite show, go for a relaxing walk and go to bed. Get up early enough in the morning to have a healthy breakfast. If you normally drink coffee, don't skip it, but if you don't regularly consume caffeine, today is not the day to start! It'll just make you jittery.

Make sure you allow plenty of time to reach the testing location and get your desk set up and ready before the examination starts.

Tips for Students with Serious Shakes

Some people don't find testing terribly anxiety-inducing. If that's you, feel free to skip this section. These tips and techniques are designed specifically for students who do struggle with serious test anxiety.

- *Control your breathing.* Taking short, fast breaths increases physical anxiety. Maintain a normal to slow breathing pattern.
- *Remember your test timing strategies.* Timing strategies can help provide you with confidence that you're staying on track.
- *Focus on one question at a time.* While you may become overwhelmed thinking about the entire test, a single question or a single passage often seems more manageable.
- *Get up and take a break.* While this should be avoided if at all possible, if you're feeling so anxious that you're concerned you will be sick, are dizzy or are feeling unwell, take a bathroom break or sharpen your pencil. Use this time to practice breathing exercises. Return to the test as soon as you're able.

Newtonian Mechanics

Units and Fundamental Constants

This section will provide an overview of the different units used in physical calculations, as well as the fundamental constants you will be expected to know. We'll also focus in on some differences between the English and metric systems, which typically cause confusion among students who are new to physics.

Mass

Mass refers to the amount of matter in an object. It is *not the weight* of an object. The common units of mass usually seen in problems are the gram, pound, kilogram, (and in some obscure physics problems, the slug). There are 454 grams in one pound, and 1000 grams in one kilogram. One kilogram equals approximately 2.2 lbs. It is a good idea to memorize these basic rates of conversion, as it will greatly aid you in your problem solving speed.

Weight

In the U.S., or imperial system, there is no differentiation in the names for the mass versus the weight of an object. However, weight is defined as a mass multiplied by a force, usually the gravitational force. For cases such as this, the weight of one pound of mass on Earth is defined as:

$$\begin{aligned} 1\,\text{lbf} &= 1\,\text{lbm} \times g_\text{n} \\ &= 1\,\text{lbm} \times 32.174049\,\tfrac{\text{ft}}{\text{s}^2} \\ &= 32.174049\,\tfrac{\text{ft-lbm}}{\text{s}^2} \end{aligned}$$

In the metric system, weight is measured in Newtons, which is equal to the force required to accelerate 1 kilogram of mass at 1 m/s². As a result, a person with a mass of 40 kilograms standing on Earth's surface, (which has a gravitational acceleration of 9.8 m/s²), possesses a weight of W = 40 kg x 9.8m/s², which is 392 Newtons.

To calculate the weight of any mass on a given surface, use the following equation:

$$W = m \times g$$

Where m is the mass of the object and g is the gravitational attractive force.

Acceleration of Gravity

The acceleration of gravity on Earth is used in many physics problems to calculate time spent in air, or a classical favorite, projectile shooting distance.

- The metric acceleration is 9.81 m/s^2
- The English acceleration is 32.2ft/sec^2

Be sure to use the correct value for the acceleration of gravity, depending on the given units in the problem!

Identifying Units

Identifying units will aid you greatly in figuring out what is missing in a problem, and how to solve it. You need to be able to recognize the name of a unit and associate it with its correct property. This will allow you to write out the equation that is necessary to solve a problem, and then see which quantity is missing. In addition, it will allow you to perform dimensional analysis, which helps to make sure that the answer you have solved for has the correct units.

In the SI/metric system, the following units are used:

Quantity	Name	Symbol
Length	Meter	m
Mass	Kilogram	kg
Time	Second	s
Electric Current	Ampere	A
Pressure	Pascal	Pa
Temperature	Kelvin	K
Amount of Substance	Mole	mol
Luminous Intensity	Candela	cd

In the English system, the following units are used:

Quantity	Name	Symbol
Length	Foot	ft.
Mass	Pound	lb.
Time	Second	s
Electric Current	Ampere	A
Pressure	Atmosphere	atm
Temperature	Fahrenheit	F
Amount of Substance	Mole	mol
Luminous Intensity	Candela	cd

Note!

- Do your best to remember measurement units by heart, alongside some of the more basic conversion factors. On the AP exam, every second counts!

Vector Math

Vectors are an integral part of the study of physics, and will show up in nearly every single type of problem. Vectors are used to show the direction of a force, whether it is a physical force in a kinematics problem, a momentum force, or a force generated by an electric or magnetic field.

A vector is defined as a quantity that involves both magnitude and direction. A quantity that has magnitude but no direction is called a scalar. For example, the value 10 m/s is a scalar value. The value 10 m/s in the northeast direction is a vector.

Vector Notation

Vectors can be identified in several ways. Usually, a vector will have an arrow above its value, as seen in the figure below:

However, this only tells us that the value is a vector, but not in which direction. A more common notation is called the unit vector notation, in which each directional value of a vector is placed into an equation:

$$\mathbf{v} = v_x \hat{i} + v_y \hat{j} + v_z \hat{k}$$

In this style of notation, there is a velocity x in the "I" direction, a velocity y in the "j" direction, and a velocity z in the "k" direction. This three dimensional vector unit notation allows us to see each of the individual forces in the vector separately. If we combine the forces and draw them out on a 3D plot, we can understand the direction of the vector. If you only want to work with 2D vectors, you can drop the last term in the equation (VzK), leaving you with the 2D form of the equation.

Below is an example of a 2D unit vector drawn out on a Cartesian coordinate map:

$$V = 4i + 5j$$

This vector has a magnitude of 4 in the "i" direction and a magnitude of 5 in the "j" direction. On the Cartesian coordinate map, this translates to a value of X = 4, and Y = 5.

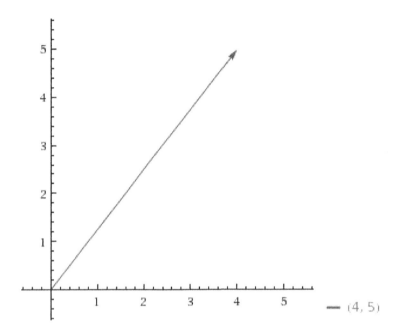

As seen above, this produces a vector with a northeast direction. When you are drawing vectors on a Cartesian plane, always make sure to put an arrowhead on the end of the vector. Otherwise, it is just a line!

Using Vectors
1. You can add and subtract vectors. Vectors are additive, such that A + B = C
2. You can multiply vectors by Scalars. Vector A x Scalar C is a possible operation.
3. You <u>cannot</u> multiply vectors by one another. The correct method of performing vector multiplication is through the dot product or cross product, not through simple multiplication.

Trigonometric Identities and Vectors

Trigonometric identities (trig identities) are conversions of trigonometric functions that are helpful for problems involving vectors and angles. Many of the problems that you will encounter involve a vector traveling at a certain angle from the horizontal. For example, a problem might state: a bullet is traveling at a 30 degree angle from the horizontal at 100 m/s. At what speed is it moving in the x-direction?

To solve this, we have to use trigonometry, and depending on the angle given, a trig identity will be needed.

Important General Formulas

You should know these trig formulas by heart:

$$\tan \theta = \frac{\sin \theta}{\cos \theta}$$

$$\cot \theta = \frac{\cos \theta}{\sin \theta}$$

$$\sin^2 \theta + \cos^2 \theta = 1$$

In addition, remember what sine and cosine mean. The sine function and cosine function are ratios of vector values to one another.

- Sine (θ) is equal to opposite over hypotenuse.
- Cos (θ) is equal to adjacent over hypotenuse.
- Tan (θ) is equal to opposite over adjacent.

What do opposite and adjacent mean? These terms refer to the opposite and adjacent sides of a triangle situated next to an angle, as demonstrated by the figure on the next page.

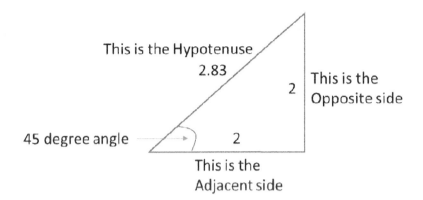

Relationships between Trigonometric Functions

<u>Opposite</u>

According to the identity, $\sin(\theta)$ should equal 2/2.83, because $\sin(\theta)$ equals the opposite over the hypotenuse sides. We check this, and we find that $\sin(\theta)$ equal 0.707, which is the same as 2/2.83!

<u>Adjacent</u>

According to the identity, $\cos(\theta)$ should also equal 2/2.83, because $\cos(\theta)$ equals the adjacent over hypotenuse sides. We check this, and find the same answer. Great!

Using Identities

According to these identities, we can easily calculate the value of the X-direction and Y-direction vectors of a given unit vector based on our knowledge of trigonometry. The sine function represents the X-direction, and the cosine function represents the Y-direction.

Sample Problem – Trigonometric Functions

A bullet is fired at 100 m/s at a 20 degree angle from the horizontal. How fast is it traveling in the X-direction?

1. First, we need to draw out a diagram of how the problem looks in real-life. On the AP physics exam, it may not be practical to spend a lot of time drawing out a diagram for each problem, so we suggest that you become familiar with the terminology of problems prior to the AP Physics exams so that you can

visualize the problem in your head, instead of drawing it out. If we draw out the problem, it would look like this:

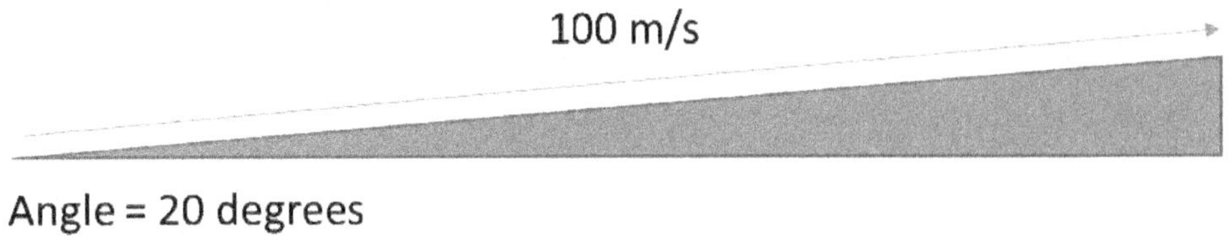

100 m/s

Angle = 20 degrees

We see that in this case, the 100 m/s vector forms the hypotenuse of a right triangle. The angle is 20 degrees, and the X-direction is the adjacent side.

2. Next, use the correct trigonometric function. As we know, the cosine function is equal to adjacent over hypotenuse. Thus, we set up the equation to be:

$$\cos(30) = \frac{x}{100 \ m/s}$$

We can then simplify this to find:

$$x = 100 \frac{m}{s} \times \cos(30)$$

This lets us solve for the x-direction velocity vector with ease, and we find that the answer is 86.6 m/s in the X-direction.

More importantly, this lets us derive a sample function that can solve for the given velocity in x with no problem:

$$x = \cos(\theta) \times A$$

Where A is any vector, so long as the angle related to the horizontal is given.

Sample Problem – Trigonometric Functions II

This next sample problem looks at using a different application of vectors and trigonometric identities. A cube with a mass of 20 kilograms is resting on an inclined plane, which has an angle of 30 degrees. What is strength of the force pulling the cube down the plane?

1. Again, we want to first draw a diagram to make sure that we understand the whole picture. We note that in this problem, there is only one force being applied: the gravitational force.

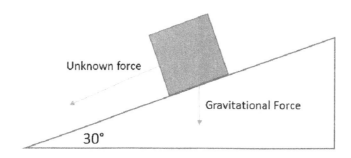

This is a bit more complicated than the first example. However, the problem is still solved simply by applying your knowledge of vectors.

2. Calculate the gravitational force. The mass is 20 kg, and the gravitational acceleration constant (metric) is 9.81 m/s². This gives us a force of 20 x 9.81, or 196 Newtons, in the Y direction.

3. Use a trig function to estimate the force pulling the block down the slope. With the angle and opposite side, we can draw a new triangle that shows the force of the block sliding down the ramp.

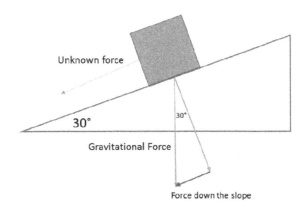

As we can see, in the new triangle, the red line represents the force coming down the slope.

Wait A Minute!

Now, you might be thinking, "Why can't we use the original triangle to solve the problem?"

The original triangle is not an adequate representation of the total forces on the mass. We have to remember that the largest force in this case is the gravitational force, and as such it should be the hypotenuse. If we used the original triangle in our calculations, we would have arrived at the conclusion that force pulling the block down the slope was 392 Newtons, almost double the gravitational force! This intuitively doesn't make sense, and as a result, we need to draw a new triangle to correctly show the forces.

Now, we use the new force triangle to set up an equation. The hypotenuse is known, and we need to know the value of the opposite side. This is a sine function, which represents opposite over hypotenuse.

$$\sin(30) = \frac{opposite}{196\ N}$$

$$\sin(30) \times 196\ N = opposite$$

Solving the equation above, we find that the force pulling the block down the slope is 98 Newtons.

Note!

The knowledge of trigonometric identities and how to apply them in physics is <u>fundamental</u> to solving the majority of problems in AP Physics B. If any part of the above section confuses you, we strongly recommend that you review this section before continuing onto the rest of the review sheet, because this is the last time in the guide you'll find detailed explanations of how vector forces and trigonometric functions are determined.

Newton's Laws

These are a set of physical laws developed by Sir Isaac Newton. They govern the interactions of forces between different objects.

First Law

Newton's first law states, "Every object in a state of uniform motion tends to remain in that state of motion unless an external force is applied to it." This is also known as the Law of Inertia.

Inertia is defined as the resistance of an object to external forces, based on existing forces. Essentially, a non-moving object is in perfect balance of forces. If an object is moving, then the forces acting upon the object are not balanced.

As a result of the first law, the concept of a normal force is brought to light. A normal force is a surface force that acts against the gravitational force. Everything that is subject to a gravitational force is also subject to the normal force if it is not moving. Here's an example involving a 1 kg paperweight resting on top of a flat desk:

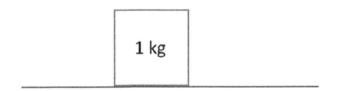

Here's a question for you: What forces are acting on this 1 kg block?

Our initial response to this question is: none! The block isn't moving. Therefore, it makes sense to assume that no forces are acting on the block. However, this isn't true. If the block is resting on a flat surface on Earth, then a gravitational force must be applied! The gravitational force in this case would be 1 kg x 9.81 m/s², or about 9.8 N.

However, according to Newton's first law, an object at rest must have a balance of forces. If the block only has a gravitational force acting on it, then it should be moving. However, it isn't moving, so there must be a force counteracting the gravitational force. This is the normal force.

The normal force acts perpendicular to the surface on which an object is resting:

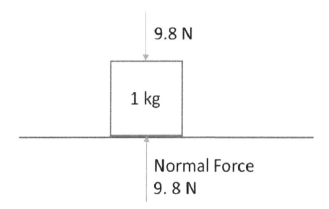

This is another foundational principle of physics. When solving problems on Earth, we always have to account for both the gravitational force and the normal force in order to obey Newton's First Law.

Second Law

Newton's second law tells us what happens when a force acts on an object that is not stationary. In this case, a movement will occur, and this is related by the second law, which states that:

$$F = ma$$

Where F is the force, m is the mass, and "a" is the acceleration that is experienced by the mass. This means that if we push on an object with a given force, we can predict how fast it will accelerate if we know the mass of the object.

Vectors

When a force is applied on an object, it must have a direction. As a result, a force is *always* a vector. Because of this, we can use trigonometric functions when working with forces.

The second law directly relates to mechanics and kinematics. These are the areas of study in physics that examine moving objects. On the exam, you may be asked questions about how fast a ball is being thrown, what amount of force is needed to move a person, or what friction force is experienced by a sliding object.

Sample Problem – The Second Law
Two people are ice skating on a rink. Assume the ice rink is frictionless. Person A weighs 40 kg. Person B weighs 55 kg. If Person A pushes person B, resulting in person B accelerating at 4 m/s², how much force did Person A use?

1. First, we need to understand all the relevant parts of the problem.

   - We know we can ignore friction in the problem
   - We know the masses of the two people. However, the mass of person A is irrelevant to the problem, because only person B is moving.
   - We are given a mass and an acceleration. From this, we can calculate the force exerted using the Second Law.

2. Now, we need to set up the equation and solve the problem. With the second law, the equation will look like:

$$F = 55 \; kg \times 4m/s^2$$

We can solve for this problem and find that F = 220 Newtons of force.

This is just a sample problem intended to demonstrate Newton's second law. On the exam, you can count on the problems being more complicated than this!

Third Law

Newton's third law is best summarized in the well-known statement, "every action has an equal and opposite reaction". Simply put, we cannot exert a force on an object without experiencing an opposite force.

The best example of this law is in the firing of a gun, which results in recoil. Although a bullet is small, with an average bullet weighing about 8 grams, the acceleration of the bullet is about 20,000 m/s².

This means that the force exerted is:

$$F = m \times a = 0.008 \, kg \times 20{,}000 \, m/s^2 = 160 \, Newtons$$

In order for the bullet to move forward, it must "push backward". This means that the gun, and whoever is holding gun, experiences a 160 N force in the opposite direction, as depicted below:

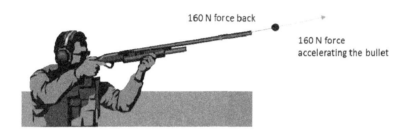

160 N is enough to knock an 80 kg (176 lb.) person back at 2 m/s^2, which is the equivalent of a strong push or shove!

Note!

Here's a cheesy mnemonic for Newton's Laws:

The first law of motion is a piece of cake

The thing won't move unless you give it a shake

The second law of motion is all about force
If you hit a thing hard, will it move? Of course.

The third law of motion gives a lot of satisfaction
Every action has an equal and opposite reaction

Sure, the poem sucks, but hey - this isn't AP English Lit review guide! Memorize this, and you won't forget Newton's laws come test time.

Kinematics and Projectile Motion

Kinematics is the study of motion. When considering motion, there are three factors:

1. Initial position
2. Velocity
3. Acceleration

These three components are related to each other through their derivatives. The use of derivatives, which is a calculus concept, is not covered in AP Physics B. However, we'll give you an explanation of the relevant equations and how they can be interpreted.

Finding Object Position

The position of an object is based on its initial location (D), in addition to a function based on the acceleration of the object and the velocity of the object. The full equation is shown below:

$$D = \frac{1}{2}at^2 + vt + d_{initial}$$

This equation states that the position of an object is a function of acceleration with respect to time, with velocity with respect to time, and to its initial position.

In the equation, "a" is acceleration, "t" is the time, and "v" is the velocity. If any of these values are zero, then the term can be eliminated from the equation. For example, if an object is moving at 2 m/s but is experiencing no acceleration, the equation would look like:

$$D = 2\frac{m}{s}t + d_{initial}$$

Sample Problem – Finding Object Position
A train is moving north at 20 m/s. It is initially located 2 kilometers west of Amityville. If it continues moving at the same speed for 5 minutes, how far away will the train be from Amityville?

1. Set up the problem and draw a small picture, if necessary. We are asked to find the location of the train from Amityville. We are given the initial location and the velocity, and told the velocity remains constant (meaning there is no acceleration).

2. Based on the above picture, we understand that if the train continues moving north, we will first have to use Pythagoras' theorem to find the hypotenuse of the triangle so we can calculate the distance from Amityville.

   To find the distance traveled, we set up the equation as:

   $$D = 20\frac{m}{s} \times 300\ seconds + 0$$

   We set the initial location to zero because this is not a straight line problem. The train is not moving in an easterly direction away from Amityville.

   Now we can solve this equation to find that the train has moved 6,000 meters, or 6 kilometers, in 5 minutes.

3. Finally, we can solve for the distance from Amityville. The x-side of the triangle is 2 km. The y-side of the triangle is 6 km. Using the Pythagorean Theorem ($A^2 + B^2 = C^2$) we find that the train is √40 km away from the town, or 6.32 kilometers.

Finding Object Velocity

The equation to find the current velocity of an object can be found in several ways. The overall equation however, is as follows:

$$V = at + v_{initial}$$

This states that the velocity of an object is equal to its initial velocity in addition to the acceleration applied over time. For example, if a car is moving at 5 m/s and accelerates at 1 m/s² for 5 seconds, then its new velocity is 10 m/s.

There is another method of determining velocity, which we'll go over, but it's important to note that this method *does not* tell you the final velocity of an object. It only tells you the average velocity over a time period.

A second method of finding the velocity of an object is to look at the distance traveled over time. In this case, the velocity is calculated as:

$$V_{average} = \frac{distance}{time}$$

This means that the average velocity over a given period of time is equal to the distance traveled over that time divided by the total time. For example, if a person runs 1 mile (5280 feet) in 5 minutes, then they were traveling at an average speed of 5280 feet/300 seconds, or 17.6 feet per second.

Now, we can't say that the runner in the situation given above was moving at 17.6 feet per second when they crossed the finish line – we can't actually know that using this method, because the runner could have been moving faster or slower at any given point while running. There may be trick questions on the AP exam that try to make you reach unsupported conclusions like this.

Finding Object Acceleration

There are two methods for finding the acceleration of an object. The first is the use of Newton's second law, F = ma. This tells us the acceleration of an object based on a force applied.

The second method is the calculation of average acceleration, which is given by the equation below:

$$A_{average} = \frac{change\ in\ velocity}{time}$$

This means that we can calculate the average acceleration over a given period of time if we know the change of velocity.

Sample Problem – Finding Object Acceleration
A train enters a tunnel traveling at 20 mph. When an observer sees the train 2 minutes later, he notices that the train is now moving at 40 mph. What was the average acceleration of the train in the tunnel?

1. First, we calculate the change in velocity. Initially, the train was moving at 20 mph. Now, it is moving at 40 mph. This is a difference of 20 mph. The time that elapsed was two minutes. We set up the problem as shown:

$$A_{average} = \frac{20\ mph}{2\ min}$$

The average acceleration is thus calculated as 10 mph/min. However, this is a very confusing set of units, and may not be a possible answer choice on the exam.

2. Instead, we should simplify all the units into feet per second:

    20 mph = 29.33 feet per second

    2 min = 120 seconds

The average acceleration was thus 0.24 ft./sec^2.

Kinematic Graphs

A common type of question on the AP Exam is based on the presentation of a graph which charts position or velocity. There are several useful tricks for solving problems such as these. First, we look at a graph of displacement over time:

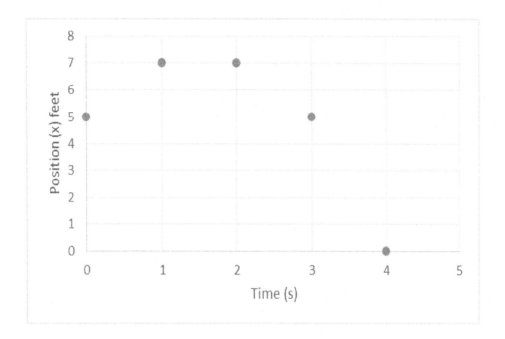

Sample Problem – Kinematic Graphs
Based on the chart above, answer the following questions: • What is the ending position of the object?   • At what time did this object have the greatest velocity?   • What is the average velocity?

1. To solve this problem, we just look at the graph. The last point on the graph is at a time position of 7 seconds. If we follow the dot to the left axis, we see that this corresponds with a position of -20 feet. Thus, the answer to question 1 is -20 feet.

2. The second question is a bit more involved. Recalling our earlier discussion on average velocity, we can calculate the average velocity between each of the time points by looking at the displacement over time. For example, from 0 seconds to 1 second, the object moved 5 feet. This means that for that period

of time, the velocity was 5 feet per second. We can then make a table reflecting all of these values, as seen below:

Time	Avg. Velocity
0 – 1	5 ft./sec
1 – 2	5 ft./sec
2 – 3	0 ft./sec
3 – 4	5 ft./sec
4 – 5	5 ft./sec
5 – 6	10 ft./sec
6 – 7	10 ft./sec

As we see from our analysis, the correct answer would be either 6 or 7 seconds. During these times, the object was moving at 10 ft. per second, which is the highest velocity in the table.

3. The average velocity can be calculated from the table shown above. We can simply sum up the various velocities, and divide by the 7 seconds of travel time to find the average velocities. This gives us:

$$V_{ave} = \frac{5+5+0+5+5+10+10}{7} = 5.71 \ ft/sec$$

Velocity Graphs

In the second type of problem, the exam will give you a graph of velocity and ask questions about acceleration, as well as total distance traveled. This type of question is usually more challenging.

Sample Problem – Velocity Graphs

Based on the chart below, answer the following questions:

- What was the highest acceleration or deceleration rate during this time period?

- What is the total distance traveled by the car over this time period?

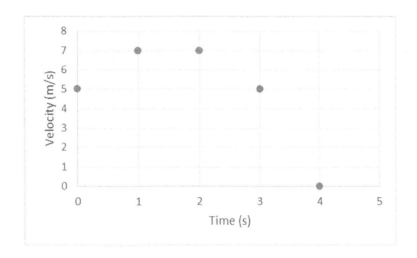

1. In order to find the acceleration rate, we remember that acceleration is equal to a change in velocity. Thus, we can find the acceleration rates in much the same manner as we calculated the velocity in the earlier problem. We take the velocity between two time periods of 1 second to find the acceleration over that time period.

   This results in the table seen below:

Time (s)	Acceleration (m/s^2)
0-1	2
1-2	0
2-3	-2
3-4	-5

   Note that in the case of velocity and acceleration, a decrease in velocity means that the car is decelerating. As a result, the answer to part one is four seconds. At four seconds, the car was decelerating at -5 m/s^2, which is the greatest acceleration or deceleration rate.

2. In the next part of the question, we have to calculate the total distance traveled. The velocity is changing, so we can't multiply the velocity by the seconds, because from zero to one seconds, the velocity changes from 5 to 7 m/s.

The best way to solve this problem is to calculate the area under the curve. The area under the curve represents the integral of the velocity component of the chart. An integral is a calculus function, and you do not need to be able to perform this operation. However, you can easily calculate the area under a curve using simple geometry.

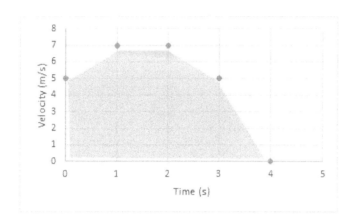

In the figure above, we want to calculate the area of the shaded green area. This will represent the distance traveled. We can do this by splitting it into four shapes:

I. Trapezoid. Height = 1. Base 1 = 5, Base 2 = 7
II. Rectangle. Base = 1. Height = 7
III. Same as shape 1.
IV. Right triangle. Base = 1, Height = 5

Solving for the areas of these shapes, we find that the areas = 6 + 7 + 6 + 2.5, for a total of 21.5 meters traveled!

Projectile Motion

Another common type of problem related to kinematics is that of projectile motion. The test will usually give an example of a projectile that is fired or launched, and ask you to solve for landing position, time in flight, or maximum height.

Sample Problem – Projectile Motion

A student throws a paper airplane at a velocity of 5 m/s off of a cliff that is 50 meters tall. They throw the airplane at an angle of 45° to the horizontal. Based on this information, and neglecting air resistance, find the answers to these three questions:

- How far will the plane travel in the x-direction away from the cliff?
- What is the greatest height that the plane reaches?
- How long will it take for the paper airplane to hit the ground?

1. First, let's sketch out a picture. This gives us the best understanding of the problem at hand. Things to include:

   I. Plane velocity and angle
   II. All forces acting on the plane (don't forget gravity!)
   III. Cliff of 50 meters

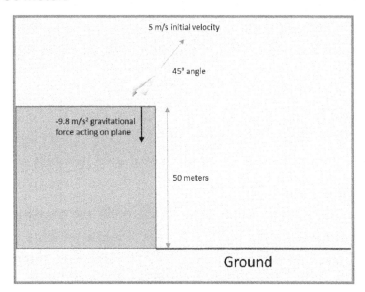

2. Now that we know what is happening in our scenario, we can get to work on solving the problem. We'll start by solving for the x and y components of velocity for the plane.

   We solve for the x velocity component to find that x = cos(45)*5 m/s = 3.53 m/s.

We solve for the y velocity component to find that y = sin*45)*5 m/s = 3.53 m/s.

Intuitively, this makes sense, because 45 degrees is halfway between the perpendicular and the horizontal.

3. Next, we solve for the maximum height reached by the plane. This is a two-step process. First, we need to find out how much time it takes for the plane to reach maximum height. Then, we can find the actual maximum height.

   Calculating the time to reach maximum height is fairly simple: We know that the y-velocity is 3.53 m/s. We know that the acceleration of gravity is -9.8m/s². This means that the y-velocity will reach zero after 3.53/9.8 seconds, or 0.36 seconds. When the y-velocity is zero, the plane has reached its maximum height, after which it will fall.

4. Now, we can use the distance equation to solve for the distance traveled. The distance equation is:

$$D = 1/2at^2 + vt + d_{initial}.$$

If we plug the information that we have into this equation, we find that:

$$D - \frac{1}{2}\left(-\frac{9.8m}{s^2}\right)(0.36s)^2 + 3.53\frac{m}{s} \times 0.36 \text{ seconds} + 50 \text{ meters} - 50.64 \text{ meters}$$

In this case, remember that the initial height is 50 meters, because we are standing on a cliff that is 50 meters above the ground.

5. Now that we know the height above the ground, we can find out how long it takes the plane to reach the ground. Again, we use the distance equation, but this time, the initial velocity is zero. This is because at the maximum height reached by the plane, the velocity of the plane is zero.

$$0 = \frac{1}{2}(-9.8)\frac{m}{s^2} \times t^2 + 50.64\ m$$

Because the initial velocity is zero, we eliminate that term from the problem. Note that if the initial velocity is not zero, it makes the question much harder to solve using just a single equation. After we do the math and rearrange the equation, we find that:

$$\frac{-50.64\ m}{-4.9\ m/s^2} = t^2$$

We can eliminate the "meters" term because they cancel each other out:

$$10.33\ s^2 = t^2$$

$$t = 3.21\ seconds$$

6. Great! Now we have our answer, right? No! This is the calculated time that it takes the paper to plane to reach the ground from <u>its maximum height</u>. The plane's total flight time is equal to this value, plus the time it took to <u>reach</u> the maximum height, which we calculated earlier to be 0.36 seconds.

   Thus, the final correct answer is 3.57 seconds of total flight time.

   Now, that we have all of these components, it's easy to solve for the distance that the paper plane travels. In the x-direction, the plane experiences no change in acceleration or velocity. Thus, it remains moving at 3.53 m/s the entire time. We now know that the plane travels for a total of 3.57 seconds. Thus, the distance equation is set up as:

   $$D = vt$$

   $$D = 3.53\ \frac{m}{s} \times 3.57\ seconds = 12.6\ m$$

   We need to ask ourselves if all of our answers make sense. We know that paper airplanes can glide very well. However, in this problem, we are ignoring air resistance, which means the plane won't glide or be slowed by air at all. In essence, the plane is thought of as a small, dense ball. So if we throw a ball off a cliff, it will take 3.57 seconds to reach the ground, and will travel about 12.6 meters away. That seems to make sense!

Note!

On the AP Exam, you can:

- Draw a picture if the problem is complicated
- Decide which part of a problem to solve first
- Include ALL forces in the problem, even ones not mentioned (like gravity and the normal force)

Remember you can calculate distance traveled by finding the area under the curve of a **velocity** graph. This doesn't work for a position or acceleration graph. Only velocity!

Friction

Friction is a force that is exerted between a surface and an object. It is a "contact force", and only exists at the interaction point between the two objects. The friction force is parallel to the surface, as indicated below:

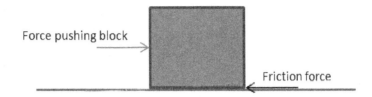

There are two types of friction: <u>static friction</u> and <u>kinetic/dynamic friction</u>. Static friction refers to a friction force that exists when an object is not moving. Kinetic friction refers to the friction force that exists when the object is moving. They are not the same!

The static friction force is usually significantly higher than the kinetic friction force. If you imagine pushing a large box across the floor, there is an initial resistance that is difficult to overcome. However, as you push the box and it starts moving, the force required to push the box appears to lessen. This is the difference between static and dynamic friction.

Friction Coefficient

A friction coefficient is a value between 0 and 1 that establishes the strength of the friction force on an object. This leads us into the equation for frictional force, which is as follows:

$$F = \mu_s \times F_n$$
$$F = \mu_k \times F_n$$

These two equations represent the calculation of friction force for the static and kinetic friction coefficients, respectively. The term Fn refers to the normal force exerted by an object. For example, if a 10 kilogram box is sitting on a surface with a static friction coefficient of 0.3, then the force required to overcome the friction is 10 x 9.81 x 0.3 = about 28 N of force.

Friction Problems

On the AP Exam, the most common type of problem involving friction is one in which a large object sits on an incline plane. You may be asked to calculate the speed at which the object slides down the plane, or whether or not the object will slide at all. Another common problem is a table and pulley problem. In this style of problem, an object sits on a table connected to a weight hanging over the edge of the table. The problem will ask whether or not the object will start to slide or not, due to the force exerted by the weight.

Sample Problem – Friction
Based on the figure and information given below, determine whether or not the block will start to slide down the hill. If the block does start to slide, what is its initial rate of acceleration?

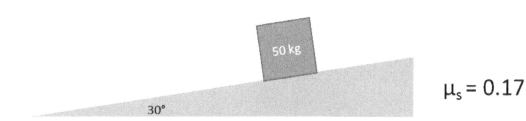

1. The first thing we should do is draw a force diagram to correctly depict the forces that are acting on the block. This force diagram will look like:

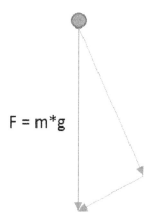

The only force acting on the block is the gravitational acceleration of 9.81 m/s². From this, we can calculate the gravitational force as being 490 N. Then, using a trigonometric function, we find that sin(30)*490 = 245 N.

Thus, we have found that there is 245 Newtons of force pulling the block down the slope. The next question that we need to answer is whether or not the friction force is strong enough to prevent the block from sliding down the slope. We are given a coefficient of static friction of 0.17.

2. Now, we calculate the normal force, which is the force that is perpendicular to the surface of the block. Thus, we can find the normal force by cos(30)*490 = 424 N of force. The equation to determine the friction force is:

$$F = \mu_s \times F_n$$

$$F = 0.17 \times 424 \, N = 72.08 \, N$$

The frictional force is determined to be 72.08 N. Because the friction force is less than the force pulling the block down the slope, the correct answer is: yes, the block will start to slide down the slope.

3. The last step is to determine the block's initial rate of acceleration. We do this by finding the difference between the friction and the accelerating forces (245 – 72 = 173 N of force accelerating the block.)

Now, we can use Newton's second law, F = ma, to calculate the initial rate of acceleration. We find that 173 = 50kg*a, and then calculate that "a" is equal to 3.46 m/s², which is equal to the initial rate of acceleration. Note that the initial rate of acceleration is much different from the average rate of acceleration when this block starts to slide down the slope. This is because once the block starts to slide, we will need the kinetic coefficient of friction to determine a new friction force, which will result in a new acceleration value that is different from the initial value calculated using the static friction coefficient.

Note!

- Static friction and kinetic/dynamic friction are not the same
- The static force is usually stronger than the kinetic/dynamic force
- The most common problem on the AP exam is similar to the sample problem in this chapter, concerning a block sliding down a slope.

Work, Energy, & Power

Work and Energy are both measures of energy in a system and have the SI unit of the joule (J). Power is work performed over time, or energy consumed over time, and is measured in watts (W). In this section, we will overview the important equations related to Work, Energy, and Power, as well as some fundamental physics concepts.

Law of Conservation of Energy

The conservation of energy is a fundamental law in physics (also known as the First Law of Thermodynamics). Simply put, it states that energy must be conserved in a system, and pure energy cannot be created or destroyed. It must come from somewhere. However, this has little practical meaning in terms of the AP Exam. For the AP Exam, the most important concept to recognize is this: If "x" amount of energy enters a system in your problem, then "x" amount of energy must leave the system or be added to the system. Here's a simple example:

Sample Problem – Conservation of Energy
Gasoline has a heating value of 18 MJ/kg. An engine consumes 2 kg of gasoline and produces 21 MJ of work. What happened to the rest of the energy?

We know that energy must be conserved in this system. An engine has consumed 2 kg of gasoline, which contains a total of 36 MJ of energy. However, only 21 MJ was used to produce work. There is 15 MJ of energy "missing". From the Law of Conservation of Energy, we know that this is not correct. Therefore, we need to propose where the rest of the energy went.

The two most common explanations is energy used for **heat** and for **sound**. If you think about it, an engine gets very hot when it is in operation, and also produces a lot of sound. These are alternate forms of energy, and are the best explanation for where the rest of the energy went.

On the AP Exam

On the AP exam, always check to make sure your answer makes sense when working with problems related to energy. You can't make more energy than entered the system, and you should be able to account for all the energy you put into a system!

Work

Work is defined in physics as force exerted over a distance. Looking at the dimensional analysis of this, we can see that:

$$\frac{kg \times m}{s^2} \times m = \frac{kg \times m^2}{s^2}$$

The units of the Joule (J) are thus kg*m^2*s^{-2}.

Work has to take place over a distance. If an object is acted upon by a force, but does not move, then no work has been performed. For example, if you push against a wall with 50 N of force, you have not performed any work. If you lift a baseball off of the ground, then you have performed work.

Work is a scalar quantity, and cannot be expressed in terms of a vector. However, the force applied to do work is a vector quantity, and thus the net force used to perform work can change depending on the angle at which it is applied. A classic example of this that often shows up on the AP exam is the use of a rope to haul a block of stone. The rope is situated at an angle to the block of stone. As a result, not all of the force that is exerted by the person is actually transformed into work. We can see this in the problem below:

Sample Problem – Work

Calculate the work performed by a person pulling a block of stone with the rope at a 45 degree angle from the horizontal. The person is exerting a force of 200 N and the block is 100 kg. He drags the block 10 meters.

1. First, we draw a simple point diagram of the forces involved. Prior to this problem, we have been drawing full diagrams, but when in a hurry, a simple point diagram will suffice.

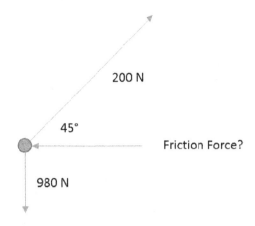

This diagram shows the gravitational force of 980 N, the force exerted by the person of 200 N at a 45 degree angle, and a possible friction force. However, the problem does not state a friction force, so we will ignore it for now.

2. Calculate the force exerted in the X-direction by the person. Because the force is applied at an angle, not all of the force is used. The amount of force is equal to 200 N * cos(45), which is 141 N. Thus, the force applied in this problem is 141 N.

3. Use the work equation to calculate the total work done. The work equation is W = F*d. The distance is 10 meters. The force is 141 Newtons.

   Thus, the answer to this problem is 1410 Joules of work was performed.

Kinetic and Potential Energy

Kinetic energy is defined as the energy of a moving object. For example, a baseball in flight has a kinetic energy. Potential energy is defined as the energy of an object that is at rest at a height. For example, if a person is standing at the top of a set of stairs, he or she has more potential energy than if they were standing at the bottom of the flight of stairs. Kinetic and potential energy also have the same units as work, in Joules.

The equation for kinetic energy is:

$$K_o = \frac{1}{2}mv^2$$

The equation for potential energy is:

$$K_p = mgh$$

Where "m" is the mass, "v" is the velocity, "g" is the gravitational acceleration, and "h" is the height of the object.

Potential energy can be transferred into kinetic energy! This is an important concept that is likely to show up on the AP Exam. Essentially, an object with a certain amount of potential energy can transfer that into kinetic energy, assuming no energy is lost through friction or heat. We will demonstrate this by looking at a sample problem.

Sample Problem – Kinetic and Potential Energy
A bowling ball with a mass of 1 kg is sitting on top of a building that is 20 meters tall. For the following questions, neglect air resistance. • What is the potential energy of the ball? • If the ball is dropped, how fast is it moving when it hits the ground? • What is the kinetic energy of the ball just before it hits the ground?

1. First, we solve for the potential energy of the ball. This part of the problem is the easiest. We simply use the equation for potential energy, Kp = mgh, and find that Kp = 1 kg * 9.81 m/s² * 20 m = 196.2 Joules. Too easy!

2. Now, we need to find out how fast the ball is moving when it hits the ground. By calculating how long it takes the ball to hit the ground, we can find out how fast it is going.

We will set up the distance equation as follows:

$$0 = \frac{1}{2}(-9.8\frac{m}{s^2})t + 20m$$

There is no velocity term because the initial velocity is zero. The starting height is 20 meters, and the final height is zero meters.

From this, we find that the time required for the ball to hit the ground is 2.02 seconds. That's a good flight time for a bowling ball!

Now that we have the time, we can find the velocity using the velocity and acceleration relationship. Again, this is similar to the projectile motion problem. This equation is set up as:

$$V = 9.8\frac{m}{s^2} \times 2.02\ s$$

We find that the velocity just before it hits the ground is 19.8 m/s. Pretty darn fast! For reference, that is equivalent to 44 mph.

3. Now that we have the velocity of the ball, we can finally calculate the kinetic energy. Remember, we have to have a correct velocity in order to determine the kinetic energy. Otherwise, we will get a wrong result.

The kinetic energy equation is set up as:

$$K_o = \frac{1}{2}1kg \times (\frac{19.8m}{s})^2$$

We solve for the kinetic energy, and find that it is equal to 196 Joules! That is pretty much the same as the potential energy. Based on this case study, we can conclude that if we drop an object from a given height, its potential energy will be directly transferred into kinetic energy.

Check your units! Work and Energy have units of $kg*m^2*s^{-2}$. If your answer doesn't have these units, you probably got it wrong. You also need to remember that potential energy and kinetic energy can be converted into one another.

Power

Lastly, in this section we will look at power. Power is force applied over time. If you recall the example from the previous section, we told you that you don't perform any work if you push hard against a wall, since the wall isn't moving. However, you are expending power.

For example, if you push against the wall for 10 seconds with 50 N of force, then your average power output for those 10 seconds was 50 watts. The equation for power is:

$$P = F \times t$$

Where P is power, F is force, and t is time. The units of power are Watts. Note that in the above example, you did not expend 500 watts. Instead, the correct terminology is you expended power at a rate of 50 watts for a period of 10 seconds. Since 1 watt per second is equal to the consumption of 1 Joule per second, you used 500 Joules of energy.

Power on the AP Physics Exam

On the exam, questions relating to power will generally ask you to compare the power consumption of two modes of action. For example, which action consumes more power?

- Moving a 10 kg block 10 meters in 5 seconds
- Moving a 10 kg block 10 meters in 10 seconds

Initially, the answer seems to be that both actions consumed the same power. After all, the work performed in both cases is the same: 100 Joules.

However, the power expended in each case is different, due to the movement speed of the block. In the first case, the average speed of the block is 2 m/s. This means that the energy of the block is 20 Joules. This 20 Joules has to be maintained over the 5 seconds. Thus, you have to expend 20 watts of power for 5 seconds to move the block.

In the second case, the kinetic energy of the block is just 5 Joules! This is due to its slower velocity of 1 m/s. This means that you only need to use 5 watts of power for 10 seconds to move the block.

Note!

- The Law of Conservation of Energy states that energy must be conserved in a system, and pure energy cannot be created or destroyed.
- Work must happen over a distance, and not all exerted force directly equates to work.
- Kinetic energy is the energy of an object in motion, and potential energy is the energy of an object at rest at a height.
- Power is force applied over time. Watts are not additive - they are a *rate* expressed in units of time, and don't directly correlate to joules of energy.

Momentum

Momentum is defined as P = mv, or the product of mass and velocity. The relationship of momentum of force is through an integral, such that the force is equal to a change in momentum over a change in time. The units of momentum are:

$$P = kg \times \frac{m}{s} = \frac{kg \times m}{s}$$

The relationship of momentum to force is as shown:

$$F = \frac{\Delta P}{\Delta t} = \frac{kg \times m}{s^2}$$

If we check the units for force, we see that it matches up with what we expect. This means that our equation to check the units is correct.

Momentum is one way of measuring the tendency of an object to continue moving in the same direction it was originally. Momentum is a conserved property in most cases. For example, if two objects collide, there should be the same momentum before and after the collision.

Impulse

The change in momentum over a specified time is called impulse. Impulse is equal to:

$$J = \Delta P$$

It is also equal to a force delivered over a period of time:

$$J = F \Delta t$$

Again, if we check the units of these two equations, we should find that they match up with the units established earlier for momentum.

On the AP Exam

On the AP exam, typical problems involving impulse will be situations involving kicking a ball or pushing briefly on an object. Here is an example:

Sample Problem – Impulse
A soccer player kicks a soccer ball with a mass of 0.5 kg. His foot is in contact with the ball for 10 milliseconds, and the ball has a velocity of 20 m/s. <ul><li>What was the impulse delivered to the ball?</li><li>What was the force the player used to kick the ball?</li></ul>

1. In this problem, we are given the mass, the velocity, and the time. We can initially calculate the momentum of the ball using the equation above. We find that the momentum is equal to 0.5 kg x 20 m/s, or 10 kg*m/s.

2. Now we need to figure out the impulse. There are two equations we can use. Which is more appropriate? Although we are given a time of contact, we are not given a force exerted. Thus, we cannot use the relationship J = FΔT.

   Instead, we will have to use the relationship J = ΔP. We know the starting momentum of the ball was 0. We know the final momentum of the ball is 10 kg*m/s. Thus, the difference of these two values, which is 10 kg*m/s, is the impulse delivered to the ball.

3. Now we can calculate the force used to kick the ball during the 10 milliseconds. This time, we can use the equation J = FΔT and rearrange it to:

$$F = \frac{J}{\Delta T} = \frac{10 kg \times m/s}{0.01\ s}$$

Thus, we find that the average force exerted was 1000 N. That's a pretty good kick!

Conservation of Momentum

Momentum is a conserved property in a collision. For example, if two objects collide with one another, the total momentum will be the same before and after the collision. However, the kinetic energy of the system will usually not remain the same. The conservation of momentum equation is given as:

$$M1V1 = M2V2$$

Essentially, this states that even though the velocities may change, the total momentum before and after a collision must remain the same. However, because the velocities of the objects may change, this means that the kinetic energy of the system is unlikely to stay the same.

Sample Problem – Conservation of Momentum
A ball with a mass of 5 kg and a velocity of 10 m/s collides with a second ball of mass 2 kg that is initially at rest. After the collision, the first ball (5 kg ball) has a velocity of 2 m/s. • What is the total momentum in the system before and after the collision? • What is the total kinetic energy in the system before and after the collision?

1. First, we will calculate the momentum and kinetic energy of the system before the collision. Only one ball is moving initially, so this is pretty straightforward.

   Momentum = 5 kg * 10m/s = 50 kg*m/s

   Kinetic energy = ½ * 5kg *(10m/s)² = 250 Joules

   How did the system's momentum change? After the collision, the first ball is moving at 2 m/s, so the first ball's momentum is now 10 kg*m/s. Momentum is conserved, which means that the second ball must have a momentum of 40 kg*m/s.

   $$40kg \times \frac{m}{s} = 2\ kg \times v$$

   Solving for the velocity, we find that the second ball should be moving at 20 m/s.

2. Now that we know the velocity, we can calculate the kinetic energy before and after the collision.

   Ball 1 Kinetic Energy = ½ 5 kg *(2m/s)² = 10 Joules

   Ball 2 Kinetic Energy = ½ 2 kg *(20m/s)² = 400 Joules

   Thus, the total kinetic energy in the system increased, showing that kinetic energy was not conserved.

Wait a Minute!

You might be wondering how the kinetic energy in the system increased, given that we just covered the Law of Conservation of Energy, which states that the energy in a system must remain constant. In this problem, we might expect to lose kinetic energy in the collision, but we certainly shouldn't gain energy. What's wrong?

In this case, we have an example of what is considered a non-real problem. Actually, if a ball with a larger mass hits a smaller ball, there is not much of a transfer of momentum. The large ball should continue moving with almost all of its speed (7 or 8 m/s in this case). The velocity of the small ball will never be greater than that of the large ball. You can picture this yourself. It doesn't make sense to hit a small object at 10 m/s, and have the small object rocket off at 20 m/s, does it?

There are sometimes trick problems like this on the AP Exam, especially on the free response section. When this happens, make sure to note down carefully the reasons for why you believe the problem is incorrect.

Completely Inelastic Collisions

A completely inelastic collision is one in which the objects stick together after the collision. In completely inelastic collisions, kinetic energy is never conserved, and the loss of kinetic energy is maximized.

Sample Problem – Completely Inelastic Collisions

A ball of silly putty with a mass of 0.1 kg and a velocity of 1 m/s strikes a baseball with a mass of 0.2 kg and a velocity of 5 m/s moving in the opposite direction.

- What is the momentum of the system after the collision?
- What is the change in kinetic energy?

1. The objects are moving in opposite directions, so we need to assign one of them a negative velocity value, because velocity (and momentum) are vector quantities.

    In this case, we will make the smaller mass of the silly putty have a negative velocity. As a result, the momentum is negative as well. Thus, the momentum prior to collision is: 0.1kg*(-1m/s) + 0.2kg*(5m/s) = 0.9 kg*m/s

2. After a completely inelastic collision, the objects are stuck together. This means that we have one mass of 0.3 kg. The momentum is conserved at 0.9kg*m/s, and thus the new velocity of the combined silly putty and baseball is 3 m/s.

3. Calculate the kinetic energy before and after. The kinetic energy before:

    ½ 0.1kg * (1m/s)^2 = 0.05 Joules

    ½ 0.2kg * (5m/s)^2 = 6.25 Joules

    Total of 6.30 Joules

    After the collision, the kinetic energy is: ½ 0.3kg *(3 m/s)^2 = 1.35 Joules

This is a loss of nearly all the kinetic energy in the system. There are two reasons for this: The collision was in opposite directions, which results in a slowing down of both masses. Also, the energy was released in the forming of bonds between the two objects, and also due to sound or heat formation.

Note that in this problem, energy was lost, but we have an explanation of why energy was lost. Energy can never be gained in a system unless there is some input (i.e. someone hits the baseball with a bat).

Note!
- Impulse is the change in momentum over time.
- Even though velocities change, the total momentum before and after a collision must remain the same.
- A completely inelastic collision is one in which the objects stick together after the collision.

Circular Motion

Circular motion is the movement of an object around a radius. However, if an object moves at a constant velocity in a circle, it must be subject to some form of acceleration, as seen below.

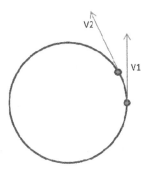

The velocity of a particle moving in a circle is always tangential to the path it is taking. However, if the velocity path is changing, and the velocity remains the same, then there must be some form of acceleration that takes place to change the vector path of the velocity. This is called the <u>centripetal acceleration force</u>. This acceleration force is responsible for changing the vector of the velocity as it moves around in a circle. If you've ever been on a merry-go-round, then you've felt the effects of centripetal acceleration as you move around and around. The equation for centripetal acceleration is:

$$a = \frac{v^2}{r}$$

Thus, it is proportional to the exponent of the object moving in a specified radius, r, around the circle.

The Accelerating Force

As we know from Newton's second law, in order for acceleration to exist, a force must be responsible. In this case, the same equation is used, but linear acceleration is replaced with centripetal acceleration to generate a new equation:

$$F = ma_c = m \times \frac{v^2}{r}$$

Using this equation, we can solve the majority of problems related to circular motion at constant velocity on the AP Physics B exam. A sample problem is shown below:

Sample Problem – Circular Motion
A child is riding on a merry-go-round, which has an outer radius of 2 meters. The child's mass is 20 kilograms, and is able to hold onto the merry go round with a force of 100 N. Based on these facts, what is the fastest speed that the merry-go-round can spin without throwing the child off?

1. We don't want the poor child to fall off, so we'd better make sure we know how fast we can spin. Fortunately, this problem is relatively simple.

We first set up the force equation with the variables that we are given. We know the child's mass and the force they are capable of using, and we know the radius of the merry-go-round. All we need to do is calculate the maximum velocity. The equation is set up as:

$$100\ N = 20\ kg \times \frac{v^2}{2\ m}$$

2. We rearrange this equation to isolate the velocity component, as shown:

$$v^2 = \frac{100N \times 2m}{20\ kg}$$

And then solving for the velocity, we find the answer is 3.16 m/s. This is actually pretty fast, so the child is in no danger of falling off even if we give her a hefty spin.

Torque

Torque is a form of work that is applied in a circular motion. If you have ever used a wrench, then you have applied torque in order to get the job done. The specific definition of torque is a force that is applied over a distance and an angle to generate circular motion, as seen on the next page.

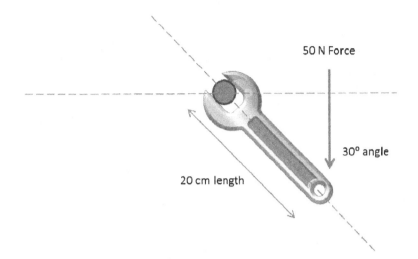

In the picture above, we see that when we turn a wrench, we are applying force at an angle, in this case 30 degrees. Furthermore, the length of the wrench increases the amount of torque that you can apply. You may have noticed in the past that the longer a wrench is, the easier it is to loosen or tighten a bolt. Based on this, the equation for torque is:

$$\tau = rF\sin(\theta)$$

Where *r* is the radius or length the force is applied on, F is the force, and theta (θ) is the angle at which the force is applied.

Torque at a 90° angle

If the torque is applied at a 90 degree angle, then sin(90) = 1 and the full force is applied. If the angle is less than 90, then part of the force goes to waste, because it is applied in the x-direction and not the y-direction. This is the reason that the sine function is used to calculate the actual force exerted over the length.

On the AP Exam

Problems involving torque on the AP exam commonly mention pulleys, wrenches, and sometimes levels. Problems using pulleys are perhaps the most common, and test your knowledge of how a pulley can be used to amplify the force of a person. The problem below incorporates many of the concepts used earlier in this review, as well as torque and a pulley.

Sample Problem – Torque

Based on the diagram below, will the mass on the table start moving, and if so, at what initial rate of acceleration?

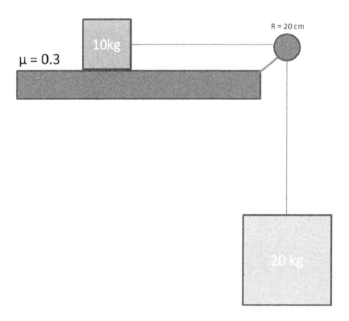

1. In this problem, we have to look at force and acceleration, friction, and also a pulley. First, we look at the diagram and come to several conclusions:

    - The blocks are connected by ropes. This means that the blocks must have both the same velocity and acceleration. This will be true as long as the rope does not break.
    - The block on the table has a countering friction force.
    - The block hanging from the pulley is applying its force across a radius of 0.2 meters.
    - The tension, T, in the rope, must be equal in both segments of the rope.

2. Now, we need to set up the equations for each block.

    Block on table:

    $$\mu M g - T = M a$$

    This equation states that the friction force minus the tension force is equal to the acceleration of the block.

    Block hanging from pulley:

    $$mg \sin(\theta) - T = ma$$

    Note that for the block hanging from the pulley, the force is applied at a 90 degree angle. This means that sin(θ) is equal to 1, and we can remove this term from the equation.

3. Now we have a problem. We don't know what the tension (T) in the rope is. However, we do know that we can add and subtract equations with similar variables. If we subtract the equations from each other, we result in:

    $$\mu M g - mg = (M - m)a$$

    Now, we have an equation that we know all the variables to. If we calculate these values, we get:

    $$294N - 196N = 10kg * a$$

    Solving for the acceleration, we find that it is 9.8m/s². This means that the block should indeed start accelerating, and it should accelerate pretty quickly. If we had found that the acceleration was negative or zero, then that means the friction force on the top block would have been too high, and the block would not have started to move.

Gravitation and Circular Motion

Let's take a brief look at the principles of gravity between two masses and the calculation of the gravitational force. On the exam, you should not expect to see more than 1 or 2 questions on this topic, if at all, but it's important that you know it.

The law of gravitation established by Newton states that a gravitational force exists between two objects, and is proportional to the masses of the two objects and the distance between the <u>center of mass</u> of the two objects. Note that it is not the distance between the surfaces of the two masses. The equation for gravitational force is:

$$F = \frac{Gm_1 m_2}{r^2}$$

The G is the gravitational constant, which is 6.67 x 10^{-11} N*m^2/kg^2.

Based on this low value of the gravitational force, we can immediately understand that gravity is a relatively weak force unless the masses are extremely large. This makes intuitive sense because day to day objects, such as your cell phone and your keys, are not attracted to each other due to gravitational force. However, objects with a very large mass, such as the Earth, do indeed have a large gravitational force.

Note!

- Centripetal acceleration force is responsible for changing the vector of the velocity as it moves around in a circle.
- Concerning torque, if the angle is less than 90, some force goes to waste, because it is applied in the x-direction and not the y-direction.
- Gravitational force is proportional to the masses of two objects and the distance between the center of mass of the two objects – *not the distance between the surfaces.*

Springs and Oscillations

An oscillation is a periodic motion that occurs in a specific frequency. Examples of objects that oscillate include pendulum clocks, springs, and the strings in musical instruments. The most basic form of oscillating motion is seen in the spring.

A spring is a coil of metal or other ductile material that, when compressed or stretched, will return to its original shape if not held by a force. The motion through which a spring returns to its equilibrium, or ground state, is an oscillation.

The equation governing the motion of a spring is known as Hooke's Law, and is given below:

$$F = -kx$$

Where F is the force applied to the spring, k is the spring constant, and x is the distance of displacement.

The greater the spring constant, the stiffer the spring. The springs which make up the suspension of a car, for example, have very high spring constants, in the range of 3000-5000 N/m, whereas the spring that is used in your ballpoint pen might have a spring constant value of 2 or 3 N/m.

Sample Problem – Springs and Oscillations

A slingshot has a spring constant of 200 N/m. The slingshot is used to fire a small rock with a mass of 20 grams. If a person draws back on the slingshot 30 cm, and the slingshot is in contact with the mass for 0.2 seconds, how fast is the rock moving when it leaves the slingshot?

This problem combines the concepts of a spring and spring constant, and also that of impulse and momentum, covered in an earlier section of the review. First, we need to calculate the force delivered by the slingshot.

1. We are given both the spring constant and the value of x, which is how far the slingshot was drawn back. In this case, we establish a negative value for x, because the spring is moving backward, as seen on the next page:

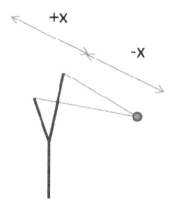

Thus, we calculate the force as F = -200N/m*-0.3meters = 60 Newtons of force produced by the spring.

2. Now we have to look at the impulse of the slingshot onto the rock. If you recall from earlier:

$$J = F\Delta t = \Delta P$$

From this, we can calculate that the impulse to be 12 kg*m/s

3. Now that we know the impulse, we can use the equation relating impulse to momentum to find the value of the velocity. The initial momentum was zero, because the rock was not moving, so final momentum is equal to the impulse.

$$12 kg \times \frac{m}{s} = m * v$$

The mass is 0.02 kg, and thus the initial velocity of the projectile is 600 m/s! That's an extremely powerful slingshot! For reference, the average bullet travels at only 320 m/s, and the speed of sound is 340 m/s. This means we have ourselves a very cool slingshot, and it'll leave behind a sonic boom when we fire a rock!

Frequency of Oscillation

When a spring is pulled and then released, it will waver back and forth until it comes to equilibrium. The rate at which this happens is the frequency of oscillation, and the time it takes to complete each frequency is known as the period. If the frequency is fast, then the time period between each oscillation is very short. If the frequency is slow, then the time period between each oscillation is very long.

The unit that expresses frequency is the hertz, which as the units of s^{-1}, or simply per second. One hertz is one oscillation per second, or in the world of computers, one hertz is one calculation per second. The basic equations that relate frequency and time period are:

$$T = \frac{1}{f}$$

$$f = \frac{1}{T}$$

Where T is the time period between each oscillation and f is the frequency. For example, if the time period between each oscillation is 0.1 seconds, then 1/0.1 = 10. The frequency is thus 10 hertz, or 10 oscillations per second.

On the AP Exam

On the AP Exam, questions relating to frequency and oscillation will ask you find the rate of oscillation or sometimes the mass of the block on a spring, given the other variables in the problem. The equations needed to solve these types of problems are:

$$f = \frac{1}{2\pi}\left(\frac{\sqrt{k}}{m}\right) \text{ and } T = 2\pi\left(\frac{\sqrt{m}}{k}\right)$$

Here, f is still frequency, pi is 3.14159, k is the spring constant, and m is the mass of the object on the spring.

Sample Problem – Frequency of Oscillation

A spring is attached to a ball with an unknown mass. A student pulls the spring back with 20N of force and releases the ball. This results in the ball moving with a frequency of 20 hertz on the spring, and the spring is displaced by 10 cm. What is the spring constant of the spring and what is the mass of the ball?

In order to solve this problem, we should draw a small diagram to depict the situation. However, in this case, it is not completely necessary, because we are given quite a few of the variables, and there are no angles in the problem.

1. We note that the spring was displaced by 10 centimeters with 20N of force. Based on Hooke's law, we can calculate the value of the spring constant (F=kx) so 20N = k*0.1m. The spring constant is 200N/m.

2. Now that we have the spring constant, we can solve for the mass of the ball by using the equations for frequency of oscillation seen above.

    We now have the spring constant, we know the value of pi, and we are given the frequency of the oscillation. The equation should thus look like:

$$20 = \frac{1}{2\pi}\left(\frac{\sqrt{200}}{m}\right)$$

If we rearrange this equation and solve for m, we discover that the mass of the ball must be 1.26 kg.

Pendulums

The last topic related to oscillations is pendulums. A pendulum is a mass hanging on a string that is swinging back and forth. In frictionless conditions and a perfect vacuum, once a pendulum starts swinging, it will never stop! Even in a non-vacuum condition, such as on the surface of Earth, pendulums with an adequately heavy mass are able to swing consistently for a long time. A simple pendulum is shown on the next page.

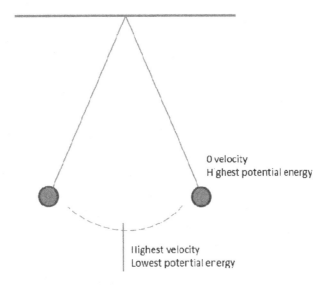

In a pendulum, as the mass swings to its lowest point while connected to the rope, the velocity is the greatest. As it reaches its height on either side, the stored potential energy is the highest, but at its peak the velocity is exactly equal to zero.

An interesting phenomenon is that as long as the mass of the object is sufficient to avoid air resistance, the frequency of the pendulum swing does not depend on the mass of the object. Basically, as long as we are not working with an extremely light object while on earth, the speed at which the pendulum swings does not depend on the mass. A 10 kg mass attached to a rope will swing at the same rate as a 20 kg mass attached to a rope. The frequency of a pendulum swing is given by:

$$f = \frac{1}{2\pi}\left(\frac{\sqrt{g}}{L}\right)$$

This equation tells us that the frequency is dependent on two things: the gravitational force and the length of the rope. Both of these make sense. The stronger the gravitational force, the stronger the force pulling the mass down, which results in a faster or slower swing. The longer the rope length, the greater the radius of the swing. If we have a 1 meter long rope, then the distance a mass must swing is significantly longer than if we have a shorter rope, and this is an inverse correlation.

Sample Problem – Pendulums

On earth, what is the frequency of a pendulum that has a rope length of 0.5 meters compared to a rope length of 1 meter?

This is a pretty straightforward problem, but we will take it one step further and make a small graph of pendulum frequency compared to pendulum length.

1. We start by plugging the values of the length into the equation given above. With this, we find that:

    Frequency of pendulum with 0.5m length rope = 0.99, or just about 1 hertz

    Frequency of pendulum with 1m length rope = 0.5

    Does this trend continue linearly? Here is a graph of the pendulum length versus frequency, assuming all other variables remain constant.

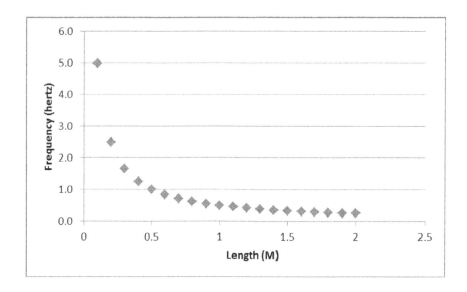

From the picture, we can see that the relationship becomes more linear the longer the pendulum becomes. However, as the pendulum becomes shorter and shorter, the frequency of oscillation becomes faster exponentially! This is a question that you might see on the AP Exam.

Note!

- Newton's second law can be used in both linear and circular motion problems.
- The speed of a pendulum swing is inversely related to the pendulum length.
- The mass of the pendulum does not have a significant effect on its period of swing, as long as we assume there is no air resistance.

Fluid Mechanics and Thermal Physics

Fluid Mechanics

On the AP Physics exam, you will be tested on your understanding of various properties related to fluids. The calculations for many of these properties is very involved and complicated, and it is unlikely for an advanced calculation problem to show up on the Physics B exam, as many of these problems are related to calculus. The core concepts in this section include density and pressure related to fluid, buoyancy of objects, flow rate through a pipe, and Bernoulli's equation, which is related to how planes are able to fly.

Density and Pressure

The density of a material is defined as its mass per unit volume. For example, water has a density of 1 gram per cubic centimeter, or 1 gram per milliliter.

The specific density or specific gravity of a material is defined as its density when compared to water. In this case, water has a set value of 1. Gold has a specific gravity of 19.3, which means that it is 19.3 times denser than water.

The pressure of a fluid is directly related to its density and to the gravitational force. The equation for pressure based on a fluid is:

$$P = \rho g h$$

Where ρ is the symbol for density, g is the gravitational constant, and h is the height of the fluid. Pressure is given in units of Pascal's, atmospheres, or bar, typically. 1 atmosphere is equal to 101.35 kPa, or 101,350 Pascal's. 1 bar is equal to 100,000 Pascal's.

The units of pressure are as follows:

$$P = \frac{N}{m^2} = \frac{kg}{m \times s^2}$$

Calculating Atmospheric Pressure

Did you know that air can be considered a fluid? As a result, the atmospheric pressure at the surface of the Earth is directly related to how much air is pressing down on us! Air has a density of about 1.2 kg/m³. The air doesn't start to really thin out until about 10-12 km into the atmosphere, which are where the highest clouds are located. However, at this point, the air density becomes much lower, around 0.5 kg/m³.

We can take an average of these values, and assuming air has a uniform density of 0.8 kg/m³, find the theoretical pressure at the surface of the earth. We use the pressure equation to get our estimate:

$$P = \frac{1.2 kg}{m^3} \times 9.8 \left(\frac{m}{s^2}\right) \times 10{,}000 \; meters = 117{,}600 \; Pascals$$

That's pretty close to the real value of 101,325 Pascals per atmosphere! Our answer is a little bit off, probably because our estimate of the density of air is a little high. If we had the correct weighted average of the density of air above us, then we should be able to get even closer to the true value.

On the AP Exam

Most of the questions that you will see involving pressure on the AP exam look at the pressure of water. If you've ever swum deep into a swimming pool, you know that the deeper you go, the more pressure you feel. One of the major questions that you will be asked is to assess which type of pool has the greatest pressure. When you look at a question like this, remember: <u>the pressure is not affected by the shape of the pool, only the depth of the pool.</u>

Which ball do you suppose experiences greater pressure?

The answer is: they both experience the same pressure! Intuitively, we might think that the ball on the left experiences more pressure, because there is more water in that tank. However, the pressure at the bottom of a pool is not correlated to the amount of total water in the tank, only the height of the water in the tank. In this case, even though the shape of the pool in the right picture is very weird and doesn't contain as much water, both balls experience the same amount of pressure.

Gauge vs. Absolute Pressure

Gauge pressure and absolute pressure are not interchangeable, and accidentally assuming they are the same thing on the AP exam may cost you points. Put simply:

$$Absolute\ pressure = gauge\ pressure + 1\ atm$$

This is true only if we are working at sea-level on earth. The gauge pressure is the pressure compared to the outside surrounding pressure. For example, a gauge pressure of zero means that there is no difference between the outside (atmospheric pressure) and what the gauge is measuring. A gauge pressure of 1 atm means that the pressure is 1 atm greater than the outside pressure. This results in an absolute pressure of 2 atm.

Sample Problem – Circular Motion

Suppose you have a closed container that is filled with water, which has a density of 1000 kg/m³.

- If the gas above the surface of the water is replaced with a complete vacuum, what is the absolute pressure experienced by an object 1 meter below the surface?

- If the gas above the surface of the water has a pressure of 5x10⁴ Pa, what is the gauge pressure of an object 0.5 meters below the surface?

To start with, we will assess the problem. In part A, the problem asks for the absolute pressure. The absolute pressure should include both the pressure attributed to the fluid, as well as the pressure above the fluid, which would be the "outside" pressure.

In part B, the problem asks for the just the gauge pressure. The gauge pressure is *only* the pressure the object resides in, or in this case, just the water. We can actually ignore the value of pressure for the gas given to us, because if we added that into the total pressure, we would be calculating the absolute pressure, and not the gauge pressure. With these items in mind, we can start solving the problem.

1. For part A, we note that there is no pressure above the surface of the water, because it is a vacuum. As a result, the pressure is solely due to the water height about the object. The height is given to us as 1 meter, so knowing this, we can perform the following calculation:

$$P = \frac{1000 kg}{m^3} \times \frac{9.8 m}{s^2} \times 1\ meter$$

    We find that the answer is 9,800 Pa, or about 0.1 atmospheres. This is the absolute pressure.

2. For part B, we ignore the pressure above the surface of the water, because we are only interested in the gauge pressure. This calculation is very straightforward. We can just replace the value of 1 meter in the above equation with a value of 0.5 meters, and arrive at:

$$P = \frac{1000 kg}{m^3} \times \frac{9.8m}{s^2} \times 0.5 \, meter$$

The pressure for part B is thus 4,900 Pa.

Buoyancy and Archimedes' Principle

Buoyancy is the ability of an object to float in a fluid that has a greater density than the object. For example, a pillow might float (briefly) in water, but a steel marble will not, due to the much higher density of the steel marble compared to the pillow.

The buoyant force of an object is a principle first determined by Archimedes, who stated that, "The strength of the buoyant force is equal to the weight of the fluid displaced by the object."

This is a very important principle, especially if you're designing boats. The total mass of the boat + carried load cannot exceed the amount of water that the hull is displacing. Otherwise, the boat will sink. For an object that floats, the fraction of the object that is submerged in the fluid is equal to the fraction of specific gravity of the object to the fluid that it is floating in.

For example, a boat has an overall specific gravity of 0.4. This means that it will float in water, which has a specific gravity of 1, and 40% of the boat will be submerged. If the boat has a height of 6 feet, then 2.4 feet of its hull will be underwater, with 3.6 feet of the hull above water.

On the AP Exam

On the AP Exam, problems will typically take place in strange environments to test your understanding of buoyant force and pressure. Problems that take place on earth with normal air and water are considered "too mundane", and the testers want to give you a challenge! Here is a problem that you might encounter.

> **Sample Problem – Buoyancy**
>
> On the moon, the gravity is about 1/4th of that on the Earth. If liquid ethanol existed on the moon, which has a specific gravity of 0.7, what would be the gauge pressure if you were submerged under 10 feet of ethanol?
>
> Furthermore, which of the following objects would float in a sea of ethanol, and why?
>
> - A brick with a specific gravity of 2
> - A piece of pumice with a specific gravity of 0.6
> - An apple with a specific gravity of 1.1
> - None of the above would float

This is a pretty lengthy problem, and as usual, we want to first consider the environment that we've been given. Gravity is ¼ of that on earth, which is about 2.45 m/s². The moon is pretty cold, so normally liquid ethanol would not exist, but we're told it does. It has a specific gravity of 0.7, which is 700 kg/m³.

1. First, we will calculate the pressure if we are submerged under 10 feet of ethanol. In order to do this calculation, we will convert the units into metric units, since the rest of the problem is in metric units. 10 feet is about 3.05 meters. Now, we can use the pressure equation to determine the pressure at a depth of 3.05 meters.

    $$P = 700 kg/m^3 * 2.45 m/s^2 * 3.05 m = 5,230 \text{ Pascals}$$

    This is quite a bit lower than the pressure would be if the same situation occurred on Earth.

2. Now, we consider the follow-up question of which of the objects would float in in the sea of ethanol. Based on Archimedes' Principle, we know that an object will only float in a material that is of greater density than itself. Based on this, the only possible choice is the piece of pumice, which has a specific gravity of 0.6, lower than the specific gravity of 0.7 of ethanol.

Fluid Flow & Bernoulli's Theorem

Based on our new understanding of fluid pressure, we can continue onto a more interesting topic: fluid flow. Fluid flow is based on the continuity equation, which states that the velocity (and volumetric flow) of a fluid versus the area through which it is flowing must remain constant. This is stated by:

$$A_1 v_1 = A_2 v_2$$

Based on this understanding, we know that if a pipe gradually becomes thinner, then the water flowing through it will become faster.

Based on this understanding, the Bernoulli Equation goes a step further, and establishes the flow rate of a fluid through a pipe depending on a series of variables:

$$P_1 + \rho g h_1 + \frac{1}{2}\rho v_1^2 = P_2 + \rho g h_2 + \frac{1}{2}\rho v_2^2$$

This equation states a relationship between the energy of a fluid flowing through a pipe, the height of the pipe, and the corresponding pressure. The best way to demonstrate the use of this equation is through a problem.

Sample Problem – Fluid Flow
A container of water is 10 meters high. It is filled with 8 meters of water, with 2 meters of air at atmospheric pressure above the water. During transport, Billy accidentally punctures a hole at the bottom of the container. How fast does the water come out of the container?

To solve this problem, we will need a picture:

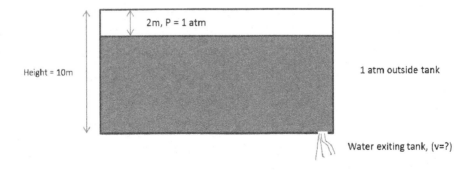

1. We see that there is 8 meters of water height in the tank. There exists a reference pressure of 1 atm both in the tank and outside the tank. Thus, in the Bernoulli equation, P_1 is equal to P_2. Inside the tank, there is no velocity. Thus, $1/2\rho v_1^2 = 0$. Finally, the height of the water goes from 8 meters to 0 meters. Based on these facts, we can simplify the Bernoulli equation to:

$$\rho g h_1 = \rho g h_2 + \frac{1}{2}\rho v_2^2$$

   P_1 and P_2 are removed because they are equal, and there is no velocity inside the tank, so the v_1 term cancels out.

2. We can also eliminate ρ from the equation because the density of water remains constant throughout the problem. Now, we can rearrange the equation further to find:

$$2(gh_1 - gh_2) = v_2^2$$

   Substituting in our known values, we find that the velocity of the water exiting the tank is: 12.5 m/s.

> # Note!
> - The density of a material is defined as its mass per unit volume.
> - Fluid pressure is affected by depth, not by shape.
> - Gauge and absolute pressure are not interchangeable.
> - On the AP Exam, problems will typically take place in strange environments to test your understanding of buoyant force and pressure.
> - The velocity (and volumetric flow) of a fluid versus the area through which it is flowing must remain constant.

Heat Transfer in Physics

Heat is defined as thermal energy that can be transmitted from one object to another. Heat is also measured in the SI unit of Joules. Heat is not an intrinsic property. We cannot say that an object *contains* 100 Joules of heat. We can, however say, that an object has *transmitted* 100 Joules of heat. The intrinsic property related to the thermal energy contained in an object is called temperature.

Temperature Scales

There are three temperature scales that you should be familiar with: Celsius, Fahrenheit, and Kelvin. Celsius and Fahrenheit are both reference temperature scales, which use water as a baseline. In the Celsius scale, 0 degrees is the freezing point of water and 100 degrees is the boiling point. In the Fahrenheit scale, 32 degrees is the freezing point of water and 212 degrees is the boiling point.

The Kelvin scale, on the other hand, is an absolute scale. At zero Kelvin, we have reached absolute zero, which is defined as the point at which no thermal energy exists in an object. We cannot have -5 K, whereas we could have -5 F or -5 C.

Conversions

Here are the following conversions for temperature that you might need to use on the test:

$$F = \frac{9}{5}C + 32$$

$$C = \frac{5}{9}(F - 32)$$

$$K = C + 273$$

From this, we know that the lowest possible temperature in Celsius is -273 C, and the lowest temperature on the Fahrenheit scale is -459 F. If a problem gives you temperatures lower than this, it is a non-real problem (trick problem!) and you should state that the problem cannot be solved.

Why is Heat Important in Physics?

Heat transfer in physics is related to phase changes, which is the change of a material from solid to liquid to gas. Also, heating a material may cause it to expand. Although these properties seem more related to chemistry, the study of phase changes and metal expansion is a vital part of mechanical engineering. On the AP Exam, typical problems might ask you about heat transfer between two objects, to interpret a phase change diagram, or to calculate the heat capacity of a material.

Heat Transfer and Heat Capacity

Heat transfer is governed by a heat transfer coefficient, which has the units of:

$$h = \frac{W}{m^2 \times K}$$

This is essentially the amount of heat energy transmitted per meter square of area per Kelvin degree. The heat transfer coefficient units tells us that a higher wattage transmitted per degree of temperature in a smaller area, the higher the heat transfer coefficient is. The equation for heat transfer through an object is:

$$h = \frac{q}{\Delta T} = \frac{(\frac{W}{m^2})}{\Delta T}$$

Where q is the heat flux through an area (watts per meter square), and the delta T is the change in temperature.

The heat capacity of an object is best defined as its resistance to temperature change. A material with a higher heat capacity will have a lower change in temperature when exposed to a greater amount of heat. For example, water has one of the highest heat capacities of 4.18 J/g*K, meaning we need 4.18 Joules to increase the temperature of one gram of water by one kelvin. On the other hand, most metals have low heat capacities. Copper has a heat capacity of just 0.39 J/g*K. This means that if we expose copper and water to the same amount of heat, copper will heat up more than 10 times faster than water will. The equation that correlates heat and heat capacity is:

$$Q = mC\Delta T$$

Where Q is the heat absorbed or released, m is the mass, C is the heat capacity, and ΔT is the change in temperature.

On the AP Exam

On the exam, you will mostly be asked heat transfer questions that require you to interpret the outcome of a situation. Questions about heat capacity will test your ability to calculate the relative heating rates of different substances.

Sample Problem – Heat Transfer I

Many houses in the United States are made from brick and wood. Brick has a heat transfer coefficient of 1.31 W/m^2*K, and wood has a heat transfer coefficient of 0.16 W/m^2*K. Alex lives in a very cold region of Canada, and is interested in building a house. Which material is best for building his house, and why?

This is a straightforward analysis and interpretation problem. Here, we are given a choice of two materials, and a question: which one should we use?

1. Based on our knowledge of heat transfer coefficients, we recognize that brick has a much higher coefficient, meaning that it will more readily transfer heat. This means that if we live in a cold region, a brick house will lose much more heat than a wood house will. For this reason, Alex should build his house out of wood in order to maximize the insulating properties of the wood, and reduce his heating bill.

Sample Problem – Heat Transfer II

A pot of water containing 750 mL of water is set on a stove. The water is initially at 20 °C. If the stove is capable of outputting 50 Watts, how long will it take to raise the temperature of the water to boiling?

This problem examines your understanding of power (Joules/sec), and heat capacity. In order to solve this problem, we will first need to calculate the total energy/heat needed to boil the water, and then solve for the amount of time it will take the stove to deliver that much heat.

1. Calculate the total heat needed to increase the temperature of water to 100°C, which is boiling.

   The equation that we will use is:

   $$Q = mC\Delta T$$

   We know the heat capacity (C) of water is 4.18 J/g*K. We know the density of water is 1 gram/mL, which gives us 750 grams of water. The temperature change is 100-20, or 80°C. Thus, the total heat required is:

   $$Q = 750g \times 4.18 \left(\frac{J}{g \times K}\right) \times 80\ K = 250{,}800\ Joules$$

2. We now know the total heat required, and we can calculate how long it will take using our heating element. By definition, one watt is one joule per second. This means every second, the stove can deliver 50 Joules of energy. We can perform the equation 250,800/50 = 5016 seconds. Then we convert this value into minutes, to find 83.6 minutes. It will take 83.6 minutes for the water boil! That might as well be forever!

Now that we have solved the problem, let's take a second to think about what this means. We know that our modern stoves can boil water in about 10 minutes, which means the stoves should be at least 8 times stronger than 50 watts. This tells us our stovetops must output at least 400W to be effective. However, this would only be true in a perfect world, where all the heat is transferred to the water. In reality, we know that at least half the heat is lost to the air, because the heating element is exposed to the air as well. This means our stovetops must be at least 800 Watts or more! A quick check online tells us that most small heating elements are 1000 Watts, and large heating elements are 1500-1800 Watts. A practical application of heat transfer in physics to the real world!

Phase Changes and Thermal Expansion

When a material is exposed to heat and its temperature increases, several things can happen. The material can melt if it's a solid. It can vaporize into a gas if it's a liquid, or if the temperature change is not too dramatic, it might expand slightly, which is true of many metal materials. As a material changes phase, an additional input of energy is needed to break the bonds in the material and allow the phase change. This energy is known as the:

- Latent Heat of Fusion: This is the energy required for a solid to turn into a liquid, and breaks the crystalline bond structures in the solid material.
- Latent Heat of Vaporization: This is the energy required for a liquid to turn into a gas, and breaks the intermolecular bonds in the liquid that hold the liquid together.

Adding latent heats <u>does not increase the temperature</u> of the material. They only facilitate the phase change. Let's look at water as an example.

- 1 kilogram of water is initially at -10°C.
- 41.8 kJ of heat will increase the temperature of the ice to 0°C
- 334 kJ of heat will now melt the ice, such that we have 1 kg of liquid water. *The temperature remains at 0°C*
- This means a total of 375.8 kJ of energy is needed to melt ice at -10°C, even though the temperature will still be 0°C.

Phase Change Diagram

A phase change diagram is a tool that can be used to understand the temperatures and pressures at which a phase change will occur for a given substance. On the AP exam, you may be presented with a phase change diagram and asked to find at which point both solid and liquid exist, or predict the phase of a substance based on the diagram. The phase change diagram of water is shown below:

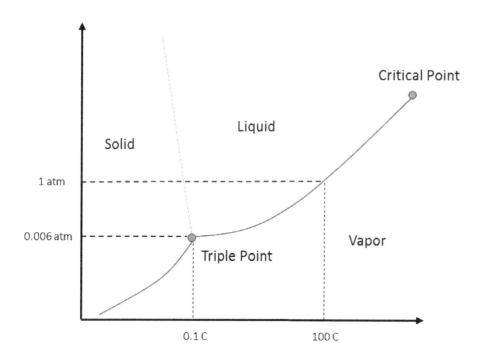

First, we note that this diagram is in log scale, not linear scale. Thus, the X and Y axes are progressing exponentially.

Next, we look at the lines that separate the different phases. There are three lines that intersect at the triple point, which represents the temperature and pressure where all three phases of water can exist! Unfortunately, due to the very low pressure required, we are unlikely to see this phenomenon in real life. The line that delineates solid and liquid is just around 0 degrees Celsius, and we can see the line that separates liquid and vapor curves up. This tells us that as the pressure increases, the boiling point of water also increases. Likewise, if the pressure decreases, we could possibly boil water at low temperatures. This phenomenon can be observed in cities with high elevation, such as Denver. Denver is located about 1 mile above sea-level, and has a correspondingly lower pressure. As a result, in Denver water boils at 95°C, which might not be hot enough to kill some bacteria or properly cook food.

On the AP exam, you may be presented with a phase diagram of a substance, such as ethanol or maybe heating oil, and asked answer a series of questions based on the diagram.

Note!

- A process where no heat enters or exits a system is called an adiabatic process.
- Fahrenheit and Celsius are reference scales, while Kelvin is an absolute scale.
- The lowest temperatures that Fahrenheit and Celsius can achieve are -459 and -273, respectively.
- On the AP Exam, questions about heat capacity will test your ability to calculate the relative heating rates of different substances.
- As pressure increases, so does the boiling point of water.

Electricity and Magnetism

Electric Point Charges

Prior to this section, our discussion has been focused primarily on mechanics and kinetics of moving objects. Here, we will discuss electric charge, Coulomb's Law, and give an introduction to some basic parts of circuits, including resistors.

Electric Charge

An electric charge is generated due to the difference in charge potential between protons, which have a +1 charge, and electrons, which have a -1 charge. Electricity is generated by the flow of electrons through a conducting coil, such as copper or steel.

The charges on a proton or electron are elementary, meaning that they cannot be further subdivided into smaller units. The charge is measured in the SI units for charge, of 1 coulomb. A single electron has a charge of -1.6×10^{-19} C. One proton has the exact same amount of charge, but of the opposite sign. A proton is a positively charged unit.

Coulomb's Law

Charged particles are naturally attracted to or repulsed from each other based on their charge. An electron will repel another electron, and a proton will repel another proton. An electron and proton will be attracted to one another strongly.

Coulomb's Law is used to predict the strength of the attractive or repulsive force between two particles. A positive value of force indicates that the particles are being repelled from one another, and a negative value indicates that the particles are being attracted. The equation for Coulomb's Law is:

$$F = k \left(\frac{q_1 q_2}{r^2} \right)$$

Where F is the electrostatic force generated, k is the proportionality constant, which depends on the material that the charges reside in, q1 and q2 represent the value of the two charges, and r is the distance between the charges.

The value of k for air is $9\times10^9 \text{N}\cdot\text{m}^2/\text{C}^2$, and this is the value that is used on most problems on the exam. If a different value is used, it will be given to you on the exam.

Based on this equation, we can calculate the force between any two given charges, as shown in the example below:

Sample Problem – Heat Transfer II
What is the force between a positive charge of 1×10^{-14} coulombs and a negative charge of 2.5×10^{-17} coulombs? The charges are 1 cm away from each other.

1. We can solve this problem by directly inputting the values that we are given into Coulomb's Law equation. This gives us:

$$F = 9\times10^9 \text{N}\cdot\text{m}^2/\text{C}^2((-1\times10^{14}\text{C})(2.5\times10^{-17})/0.01\text{m}^2) = -2.25\times10^{-17}\text{ N}$$

We can see that this force is extremely small. However, considering the size of a molecule that has this amount of charge, a 1 cm distance is actually very extreme. If we consider a more realistic distance of 10 micrometers, we find that the force generated becomes -2.25×10^{-10} N, a much more sizeable, attractive force.

Resistors and Introduction to Circuits

An electrical circuit is made of two primary components: a current source and wiring. These are the two basic components required for a circuit to exist. In addition to these two primary elements, many circuits will contain resistors, capacitors, inductors, or gates. The symbols for these elements are shown in the figure below. On the AP Physics B exam, the only circuit elements that you will need to be familiar with are a DC voltage source, a resistor, and a capacitor. You will also need to know what an inductor and diode are used for, but you will not need to perform any calculations for these circuit elements.

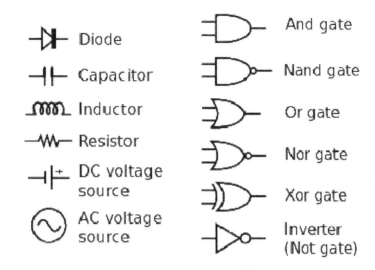

DC Voltage

A DC voltage source is short for a direct current voltage source. This is different from an alternating current in that the flow of electric charge is only in a single direction. Alternating current circuits have periodic switches of electric flow in opposing directions. AC is usually used to deliver power over long distances. The power lines that you see in a city are transmitting electric power through an AC current. However, all of the electric outlets in your house have had the AC current converted to DC.

In the DC voltage source symbol, which is represented by two parallel lines, one of which is shorter than the other, the current flows in from positive to negative. This means that if we draw a DC symbol, we can tell the current is flowing in a particular direction, as shown below.

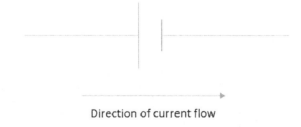

Direction of current flow

Resistors

A resistor is an electrical component that provides resistance to the flow of current through an electric circuit. A resistor is usually composed of a series of materials that are not conducive to electron flow. The units of resistance are Ohms. According to Ohm's law, the equation that relates current and resistance is:

$$I = \frac{V}{R}$$

Where I is the current, V is the voltage, and R is the resistance. Voltage is defined as the potential difference across a circuit, or a tendency for current to flow. You can think of voltage as "force" in mechanics, and current as "power". From this perspective, resistance is friction.

The above equation demonstrates to us that if the voltage is kept constant, a greater resistance will result in a decrease of current flowing through the circuit. If we draw a graph of this correlation with a constant voltage of 1.5 volts (that of a standard battery), we see that:

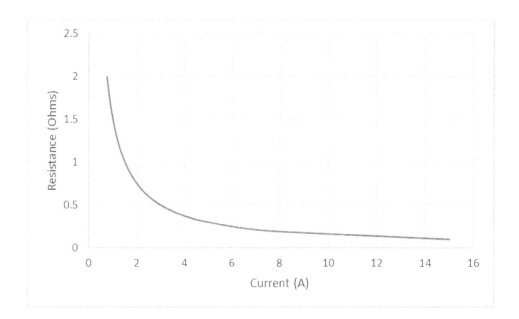

This shows us that as resistance decreases, current will exponentially increase. As resistance increases beyond a certain point, the current flowing through the resistor will approach a linear correlation. Thus, we see that decreasing the resistance from 0.2 to 0.1 has a much larger effect on current than increasing the resistance from 1 to 2.

Resistors in Parallel and in Series

Resistors in Parallel and in series have a "net resistance" that can be calculated by the following equations:

Series:

$$Rtotal = R1 + R2 + R3 \ldots$$

Parallel:

$$Rtotal = \frac{1}{R1} + \frac{1}{R2} + \frac{1}{R3} \ldots$$

This makes sense if we think about it. If we have two resistors in parallel, then the current is given two paths to flow, which will decrease the total resistance. If the resistors are in series, then there is still just one path for the current to flow and the current must now go through every single resistor, resulting in a total higher resistance.

Capacitors

A capacitor is a set of two parallel plates that carry equal but opposite charge. A capacitor is a form of storage device for electrical potential energy. When a capacitor is fully charged, no more current can flow through the capacitor until the capacitor has discharged.

A capacitor stores charge through the generation of an electric field, which creates a capacitance, measured in Farads (F). However, how does it store energy? We can think of a capacitor in terms of a water pipe to get a better understanding.

If water is flowing through a pipe, we can think of a capacitor as a rubber seal inside the pipe. The electric current we will think of as the water flowing through the pipe. As the current begins in the pipe, the capacitor/rubber seal becomes stretched. At its maximum stretching point, the capacitor can no longer "stretch", which prevents current from continuing to move in the pipe. However, because the rubber is stretched, it is storing some amount of potential energy. Once the capacitor "membrane" snaps back to its original position, then it releases the potential energy in the form of a charge.

Initial State

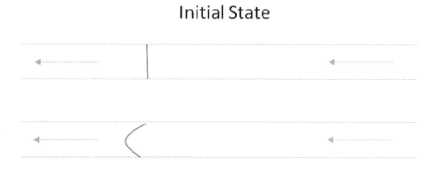

Charged State

The equation for capacitance is as follows:

$$C = \frac{Q}{V}$$

Where C is the capacitance, Q is the charge, and V is the voltage. The discharge of energy from a capacitor is given by the following equation:

$$V = V_o e^{-(t/RC)}$$

Where V is the voltage, t is the time after discharge, R is the resistance of the circuit, and C is the capacitance.

Capacitors in Series and in Parallel

Capacitors in series and in parallel behave oppositely from resistors in series and in parallel. Capacitors in series will have a total capacitance that is less than each individual capacitor, given by:

$$Ctotal = \frac{1}{C1} + \frac{1}{C2} + \frac{1}{C3} ...$$

If we think about this, it makes sense. Once the first capacitor is charged, then current through that part of the circuit must stop until the capacitor is discharged. If we look at this from the water in a pipe analogy, it becomes clear that each capacitor interferes with the ability of another capacitor in series to do its job. However, if the capacitors are in parallel, as seen below, no such interference exists.

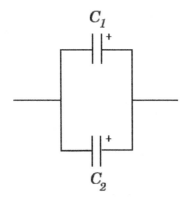

As a result, capacitors in parallel are additive in their capacitance, such that:

$$Ctotal = C1 + C2 + C3$$

Sample Problem – Capacitors
A student sets up a circuit that has two capacitors in series. The capacitances of each capacitor are 5 µF and 10 µF, respectively. The voltage in the system is 5 volts with an internal resistance of 10 Ohms, and the current is 0.2 amps. Based on this, answer the following questions: • What is the total capacitance in this system? • 50 milliseconds after the capacitors discharge, how much voltage is being discharged by the first capacitor?

The first thing we should do is draw a simple diagram that shows this system. The diagram is shown on the next page:

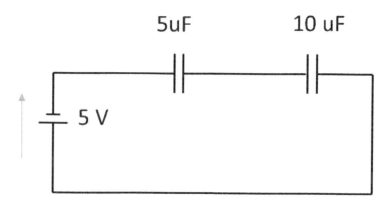

There isn't any resistance given in the form of a resistor, but we are told the internal resistance of the circuit, which the resistance of the voltage source, is 10 Ohms. That is enough to work with.

1. First of all, we will calculate the overall capacitance. The capacitors are in series so we can calculate the capacitance by performing:

$$C = 1/5 + 1/10 = 0.3 \text{ μF}$$

   As we see, the total capacitance is much lower than either individual capacitor by itself. This is as expected.

2. Now, we have to calculate the voltage being discharged by the first capacitor. This is a bit tricky. Remember, capacitors in series must have the same charge and charge potential. Because they are in series, even if one capacitor has a greater capacitance, its ability to store charge (and thus voltage) will be limited by the other.

   Thus, we cannot calculate an individual charge for the first capacitor. We will calculate the voltage for the series of capacitors and assume that the voltage produced by both capacitors is the same (which is fair assumption).

   Based on this, the equation that we will use is:

$$V = V_o e^{-(t/RC)}$$

   We have the initial voltage of 5 volts, we have the total capacitance, which is 0.3 microfarads, which is also 3×10^{-7} Farads, the time is 50 milliseconds, which is 0.05 seconds, and we have the internal resistance of the circuit, which is 10 Ohms. Based on this, the final equation will look like:

$$V = 5V \times e^{-\left(\frac{0.05}{0.000005}\right)}$$

This gives us an answer of ZERO volts! The capacitors have completely discharged at this point. If we redo the problem with a 50 **microsecond** time, we see that about half of the voltage is left. This tells us that usually, capacitors discharge extremely quickly. This also tells us that if we touch a charged capacitor, we will definitely get zapped before we can let go. The average human reflex time is between 300-500 milliseconds, but a capacitor will release its charge into you 1000 times faster than you can react.

Note!

- The value of k for air is $9 \times 10^9 N \ast m^2/C^2$, and this is the value that is used on most problems on the exam
- On the AP Physics B exam, the only circuit elements that you will need to be familiar with are a DC voltage source, a resistor, and a capacitor. You will also need to know what inductors and diodes are used for.
- Current and resistance are inversely and exponentially related.
- Current in a circuit can be thought of as analogous to water in a pipe.
- Resistors in series are additive. Resistors in parallel have a lower total resistance than any individual resistor.
- Capacitors in parallel are additive. Capacitors in series impede each other's function, resulting in a total capacitance that is lower than any individual capacitor.

Direct Current (DC) Circuit Analysis

Direct currents are the most common type of circuit, and you may be asked to assess the function of circuit, including the power being generated by the circuit and more importantly, to use Kirchhoff's rules to predict the voltage or charge across a single operation in a circuit.

Electric Circuits

In an electric circuit, a closed loop is formed that is connected to the positive and negative ends of a voltage source. The voltage source provides the charge potential that drives the current across the circuit. The power that is produced by a circuit is equal to the drop in voltage multiplied by the current in the circuit, according to the following equation:

$$P = IV$$

Remember, power is given in watts. Thus, a car battery, which has a voltage of 12 volts and a current of 200 Amps, is able to produce a power of 2400 watts. Car batteries need to be able to provide this much power in order to start the car and run electronics. When a battery becomes old, the voltage that is able to be maintained by the battery is reduced, which results in a lower power output. For this reason, car batteries that are very old are no longer able to produce the power required to start the engine, and need to be replaced.

Consumption of Power by Resistance

Resistance is present in every single conductor. The best conducting metals, such as copper and aluminum, have a very low resistance. Regardless, current traveling through these wires will still experience some resistance, and will lose power and produce heat as a byproduct. The heat produced by a resistive element is calculated by:

$$P = \frac{V^2}{R}$$

This means that all of our electronics are not 100% efficient. Whatever amount of power we put into an electronic toy or computer, we will notice that some heat is generated, which is a result of a loss of power through resistive elements.

Kirchhoff's Rules

Gustav Kirchhoff developed a set of rules in the 1800's that allows us to exam individual pieces of a circuit.

Loop Rule

The sum of the potential differences that traverse any closed loop must be equal to zero. This means that in any circuit, the voltages must sum to zero from the beginning to the end of the loop.

Junction Rule

The total current that enters a particular junction must equal the total current that leaves that junction. Similar to our analogy of water flow in a pipe, we understand that if we put water in a pipe, it must come out. The same is true of electric current.

Note!

- On the exam, you may be asked to assess the function of circuit, including the power being generated by the circuit and more importantly, to use Kirchhoff's rules to predict the voltage or charge across a single operation in a circuit

Magnetic Fields

Magnetic fields can be produced by moving electric charges inside an electric field. The SI unit for magnetism is the Tesla, named after the Russian inventor Nikola Tesla, who performed a lot of pioneering work in Magnetism. Magnetic fields can be quantified by the force that it exerts on a moving particle that is charged. The Lorentz Force Law states that:

$$F = q(E + v \times B)$$

Where F is the force exerted on the particle, q is the charge of the particle, E is the electric field in which the particle is moving, v is the velocity of the particle, and B is the strength of the electric field.

Note: You will not be asked to solve the Lorentz Force Law on the multiple choice portion of the AP exam, as the calculation involves a cross product. You may be asked to set up (but not fully solve) a magnetism problem in the free response section.

Right Hand Rule and Magnetic Fields

The force vector of a magnetic field is determined by the right hand rule, shown below:

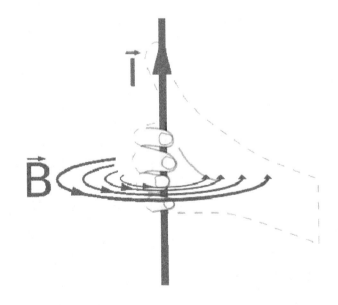

Using the right hand rule, we can determine the direction of the magnetic field as well as the force vector. The magnetic fields will curl in the direction of your fingers, and the force produced by the electric field is perpendicular to the vector of the electric field. Note that because the vector of the magnetic field is not linear, the direction of force will change depending on where the particle is located in the field.

North and South Poles of Magnets

Much like the Earth, which has both a magnetic field and poles, all smaller magnets also have poles. Magnetic field lines always flow from North to south, as demonstrated by the picture below.

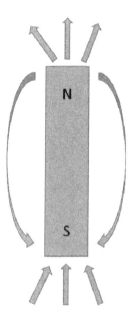

The key fact that you need to remember is that the magnetic field always extends away from the North Pole, and is attracted toward the South Pole.

If two north poles come into contact, the magnetic force repulsion between the two poles grows exponentially, such that for a strong magnetic it is nearly impossible to touch two of the same pole together. Likewise, the repulsive force also exists for the South Pole.

On the AP Exam:

On the AP Exam, you will be asked to solve problems involving charge moving through a field, to demonstrate your understanding of the direction of force of a magnetic field, and also the direction of the magnetic field based on the current flow through a wire. An example problem is given below.

Sample Problem – Magnetic Fields

A point charge that is positive charged with 3.5 x 10⁻⁶ C is moving at a speed of 3000 m/s through a magnetic field. The field has a strength of 0.25 Tesla. In addition, the particle is moving at a 30 degree angle from the horizontal. What is the force experiment by this particle?

This problem has quite a lot of elements, and in order to understand the direction with respect to the angle given, we need to draw a picture. The picture should look like:

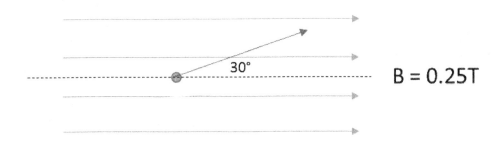

This picture shows the particle correctly moving at a 30 degree angle to the horizontal. Thus, the particle is not moving with the magnetic field.

1. Now, we can start solving the problem by using the force equation given above:

$$F = qvB \cdot \sin\theta$$

We insert the known values that we are given, and obtain:

$$F = 3.5 \times 10^{-6} C \cdot 3000 \text{ m/s} \cdot 0.25T \cdot \sin(30) = 0.0013 \text{ N}$$

The answer is thus 0.0013 N, which isn't a very large force.

Note!

- We can determine the force of a particle moving through a magnetic field by using the equation $F = qvB \cdot \sin\theta$. Furthermore, the direction of this force can be determined by using the right-hand rule.

- The force exerted by a magnetic field is always perpendicular to the direction of the velocity that a given particle is moving at. As a result, a particle that experiences a force generated by a magnetic field will almost always move in a circular motion.

- The magnetic field always extends away from the North Pole, and is attracted toward the South Pole.

- Wires that carry a linear current will also generate a magnetic field. The direction of the magnetic field is determined by the right hand rule, where the fingers curling around the direction of the current show the direction of the magnetic field. The strength of the magnetic field induced by a current is proportional to the strength of the current, and inversely proportional to the radius of the wire carrying the current.

- You will not be asked to solve the Lorentz Force Law on the multiple choice portion of the AP exam, as the calculation involves a cross product. You may be asked to set up (but not fully solve) a magnetism problem in the free response section.

Waves and Optics

Traveling Waves and Sound

A wave is defined as a recurring periodic motion. The simplest examples of waves are seen in the sine and cosine function, demonstrated in the figure below:

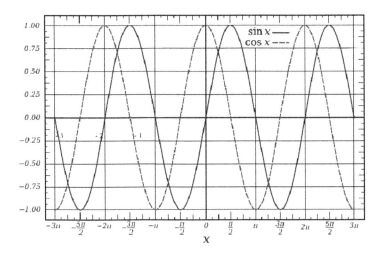

Waves are seen in physical objects every day. If you've ever snapped a rope, you've noticed that a wave travels along the rope until it reaches the other end. Other non-physical objects that have wave forms include sound, light, and radio waves.

Characteristics of Waves

The four major characteristics of waves are the wavelength, amplitude, period, and frequency:

- Wavelength: This is the distance from the peak of one wave to the next. This is measured in standard length units, usually in nanometers or micrometers.
- Amplitude: The amplitude of a wave is the distance from the top of the wave to the bottom of the wave.
- Period: The period of a wave is the time it takes to complete one complete oscillation of the wave. In the case of the sine wave, the period is the time it takes to go from 0 to 2π, which is a full oscillation.
- Frequency: The frequency of a wave is the number of oscillations that occur per second. The frequency has an inverse relationship with the period. The shorter the period, the greater the frequency.

Transverse Wave

A traveling wave, or transverse wave, can be described mathematically by the following formula:

$$y = A \sin\left(\frac{2\pi}{\lambda}\right)(vt \mp x)$$

This equation allows us to predict the y coordinate of a traveling wave based on its frequency, amplitude, and velocity with respect to time. On the AP Physics Exam, you may be asked to write the equation of a transverse wave based on this formula.

Sample Problem – Transverse Waves
What is the equation for a transverse wave that has a period of 0.01 seconds, a velocity of 2 m/s, amplitude of 3 cm, at the point where x =0?

We first need to convert the period into frequency.

2. Frequency = 1/period, thus the frequency is 100 Hertz. The velocity is given, and the x value is zero. Thus, the equation will be:

$$Y = 0.03\sin(2\pi/100)(2m/s)(t)$$

Sound Waves

Sound waves are high frequency (comparatively high frequency) waves that are usually produced by the vibration of an object, such as the strings of an instrument, or even the metal vibration in a cymbal. The vibrations cause the air to vibrate as well, which creates a pressure variation in the air. When we hear a sound, it is due to the detection of the pressure variation by our ear drums. Sound waves can also be conducted through other materials that are not air.

The speed of a sound wave depends on the medium that it is moving through. In air, sound has a speed of about 340 m/s, or 761 miles per hour.

The equation that is used to calculate the speed of sound is:

$$v = \sqrt{\frac{B}{\rho}}$$

Where B is the bulk modulus value of the medium it is traveling through, and p is the density of the material that it is traveling through. The bulk modulus of air is 1.42×10^5 Pa and the bulk modulus of water is 2.2×10^9 Pa. The greatly increased bulk modulus of water results in a speed of sound in water of about 1500 m/s, more than 4 times that of the air speed. For this reason, sound carries over and through water much better than in air!

Sound waves follow the same general equations as other waves, with the velocity, wavelength, and frequency of the sound wave being related by the following equation:

$$\lambda = \frac{v}{f}$$

Where lambda is the wavelength, v is the velocity, and f is the frequency.

Sound waves with a lower frequency have a "lower" pitch to human ears and sound waves with a high frequency have a "high" pitch to human ears. Some examples of high-frequency (treble) instruments include flutes, trumpets and violins. Typically low-frequency (bass) instruments include tubas, trombones and bass guitars. Humans can typically detect sound waves between 20 Hz to 20,000 Hz.

Sample Problem – Sound Waves
The "bass" sound of music typically varies between 300 Hz to 1000 Hz, and the "treble" sound of music typically vary between 5000 to 10000 Hz. Pick two frequencies, one of each bass and treble, and demonstrate their differences in wavelength based on the same travel speed.

As we know, sound travels at the same speed. If we are standing some distance away from a speaker, we do not hear one frequency of sound before another. This means that the velocity of sound must be constant, and instead the wavelength will change, depending on the sound. We know the velocity of sound is 340 m/s in air. Thus, we will solve this problem by selecting two frequencies to use and answering the problem.

1. We will select the frequencies of 500 Hz and 5000 Hz. We have selected these frequencies to get a good comparison, since one is 10 times the other. Using the equation, we can now solve for the wavelength of each frequency.

$$\lambda = 340/500$$

$$\lambda = 340/5000$$

The wavelength of the 500 Hz sound is found to be 0.68 meters, and the wavelength of the 5000 Hz sound is found to be 0.068 meters, or 6.8 cm. The difference in wavelength is quite significant, and is one of the reasons "bass" sounds appear to be stronger than treble sounds. The power or energy carried by a sound wave is dependent on its wavelength.

Sound Intensity

Sound intensity is measured by the units W/m^2, and is used to demonstrate the amount of energy carried by the sound wave; the louder the sound (and larger the amplitude), the greater the intensity. However, sound wave energy is inversely proportional to the distance from which it is heard. Thus, as you move away from a sound source, the intensity of the sound decreases exponentially.

Sound loudness is commonly measured by decibels, which is calculated by the following equation:

$$DB = 10 \log \left(\frac{I}{I_0}\right)$$

I_0 is the smallest intensity that can be heard by the human ear, which is 1×10^{-12} W/m^2. Any sound with a lower intensity is not discernable.

Sample Problem – Sound Intensity
Alex is watching a military jet take off. However, the staff at the air force facility make him put on earmuffs, because the sound from the jet is extremely loud. They tell him the sound intensity is about 2.5 W/m². What is the decibel level of this sound?

1. We can solve this problem by using the decibel equation, and entering the values that we are given. We already know the I0 value to be standard at 1 x 10⁻¹² W/m.

$$DB = 10 \log \left(\frac{2.5}{1 \times 10^{-12}} \right)$$

We find the decibel level is 124 decibels, which is extremely high!

Decibel Scale

The decibel scale is a logarithmic scale of sound. A sound of 20 decibels is 10 times louder than a sound of 10 decibels. On the AP exam, you need to recall that it is a logarithmic scale, and not a linear scale. Thus, a sound of 100 decibels is actually 1000 times louder than a sound of 70 decibels. It is not 42% louder. A small example of the decibel scale is given below:

Decibel Level	Example
10	Breathing
20	Whisper, paper rustling
50	Average home, speaking voice
70	Vacuum cleaner
100	Lawnmower or un-muffled car
120	Aircraft takeoff, sonic boom

Note!

- Sound wave speed is dependent on the medium it travels through. Sound waves travel much faster in water than in air.
- The amplitude of a wave (height of wave) is directly correlated to the amount of energy the wave carries. A greater amplitude equals a greater energy.
- The decibel scale, which measures sound intensity, is a logarithmic scale. It is <u>not linear.</u>

Light Refraction and Optics

Light is a form of electromagnetic radiation that makes up a small spectrum of the all electromagnetic waves. Visible light is the wavelength range of 400 to 800 nanometers, and is one of the few types of radiation that is able to penetrate the Earth's atmosphere, as seen in the figure below.

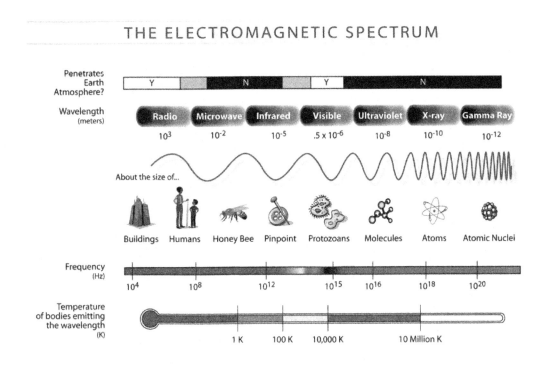

In addition to visible light, radio waves and infrared radiation are able to penetrate the earth's atmosphere and ozone layer. Ultraviolet radiation is as well, to some extent. Fortunately, the more harmful and high energy radiation types from the sun, including X-rays and gamma rays, are unable to penetrate the atmosphere!

On the AP Physics exam, the following areas are the most important to understand in light and optics: nature of light as a wave, reflection and refraction of light, and lenses.

Young's Double Slit Experiment

Prior to this experiment, light was thought of as a particle. However, the double slit experiment demonstrated that light has both wave characteristics <u>and</u> particle characteristics. To demonstrate this, Thomas Young set up a double slit experiment that showed that light had wave characteristics.

In the double slit experiment, light is shined initially through a single slit, followed by two slits. If light had purely particle properties, then the result should show two thin beams of light as a result, as seen below:

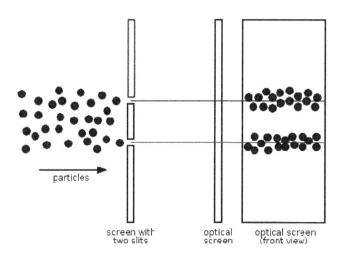

However, when Young performed his experiment, the result was actually a series of bands of light, which were proposed to have been created by the interference of light waves. Only waves exhibit interference (both constructive and destructive) properties. Thus, based on this experiment, young concluded that light could not be purely a particle, but must have wave properties as well.

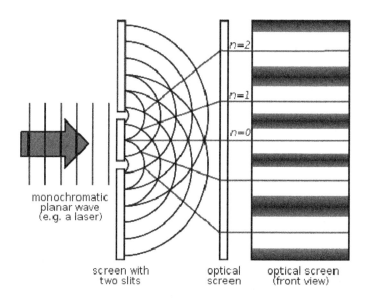

> **Note!**
> - On the AP Physics Exam, you will likely be asked to explain why light is considered to have both wave and particle properties. The best explanation is to use Young's experiment as an example.
> - You may also be asked to calculate the wavelength of light or the frequency of light given specific variables. Since light has wave properties, it follows the same equations for calculation of frequency and wavelength as sound, namely using this equation:
>
> $$\lambda = \frac{v}{f}$$

Reflection and Refraction

What is the difference between reflection and refraction? Reflection of light can be performed by almost any surface, and results in partial or total "bouncing" of light from the surface in a different direction. Refraction occurs when light passes through a substance that changes the angle at which the light exits that substance.

Plane Mirrors

A plane mirror is a mirror that is flat, which includes the majority of mirrors. When light strikes a plane mirror, it is reflected directly off of the surface at whatever the incident angle of the light was. The incident angle refers to the angle of light from the horizontal that strikes the mirror, as shown below:

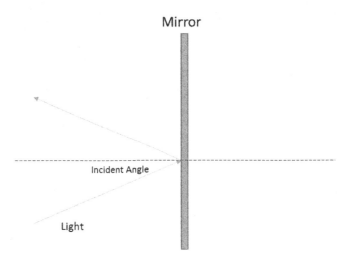

When considering a reflection from a mirror, there are four questions you should ask.

1. Where is the image located? Is it in front of the mirror or behind the mirror?
2. Is the image real? A real image is one in which light rays focus at the image. Because flat mirrors result in no focus on light, an image produced by a flat mirror is non-real.
3. Is the image upright or inverted? Flat mirrors produce upright images that are not upside down. However, depending on the shape of the mirror (such as if it is concave or otherwise), the mirror may produce a different orientation of image.

4. What is the height of the image compared to the real life object? The farther away we stand from a flat mirror, the smaller the image looks. However, because a flat mirror reflects light linearly, the image is not distorted. If we were standing an equal distance away from a real life object, it would appear the same size as if the object were placed that distance away from a mirror. Thus, for flat mirrors, there is no magnification.

Concave and Convex Mirrors

Concave and convex mirrors are also known as spherical mirrors, because the curvature of the mirror is similar to that of a sphere. A concave mirror has a surface that is curved **in** toward the center of the mirror. A convex mirror has a surface that is curved **out** away from the center of the mirror. You can think of a concave mirror as "caving" in to remember the correct direction of curvature. A picture of a convex versus concave mirror is shown below.

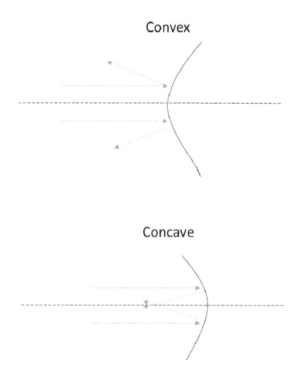

Convex mirrors result in light curving away from the centerline of the mirror, which has the effect of making the image "bulge". Concave mirrors have the effect of reflecting light toward the center line of the mirror, which makes the image appear to shrink in the middle.

There are two relevant equations that can be used to calculate the magnification of an image in a mirror, and also to determine the focal length of the mirror. The first equation is called the mirror equation, seen below:

$$\frac{1}{S_0} + \frac{1}{S_1} = \frac{1}{f}$$

Here, S0 is the objects distance away from the mirror, and S1 is the projected image's distance from the mirror. The sum of the inverse of these two values is equal to the focal length of the mirror. In this equation the value of S0 should always be positive. S1 can be either positive or negative, depending on where the mirror projects the image.

The second relevant equation is called the magnification equation, seen below:

$$M = \frac{h1}{h0} = -\frac{S1}{S0}$$

Where h1 is the height of the projected image by the mirror, and h0 is the actual height of the object reflected by the mirror. It is possible for there to be negative magnification, as in the case of a concave mirror, where the height of the projected image is less than the height of the actual object.

Sample Problem – Concave and Convex Mirrors

A room has a convex mirror that has a focal length of -30 cm. If an object that is 10 cm in height is placed 20 cm in front of the mirror. What is the height of the projected image?

1. In order to solve this problem, we first need to calculate the distance of the projected image. We know the original distance S0 =20 cm. We know the focal length as well. Thus, we can place these known values into the mirror equation and solve for S1.

$$\frac{1}{20} + \frac{1}{S1} = -\frac{1}{30}$$

We solve for S1 and find that the value is -12 cm. Thus, the projected image is actually behind the mirror.

2. Now that we know the value of S1, and S0, we can calculate the magnification of the mirror and solve for the height. The magnification is equal to:

$$M = \frac{h1}{h0} = -\frac{S1}{S0}$$

Given the values of S1 and S0, we calculate the magnification to be 0.6. This means that the height of the projected image is 60% of the original height of the object. Since the original height is 10 cm, then we find that the projected image has a height of 6 cm.

Note!

On the AP Exam, questions about mirrors will focus on the height of a projected image, whether or not the image is real or virtual, and whether it is upright or inverted.

Lenses

A lens is an object that refracts light. This means that it allows light to pass through it, but changes the path of the light or the focal point of the light. A convergent lens will converge incoming light to a focal point on the far side of the lens. Convergent lenses are used for people who are far-sighted, and cannot see objects clearly from short distances.

A diverging lens causes incoming light to diverge away from the centerline of the lens. This creates a focus that is actually in front of the lens. However, light that passes through the lens is refracted away from a focal point. A diagram of these two lens types is seen below:

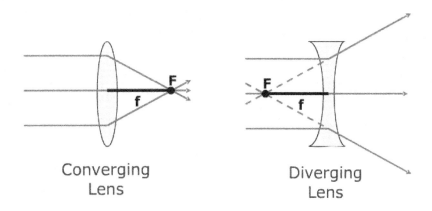

Converging Lens   Diverging Lens

The equations for the refractive index and magnification of a lens are the same as that for mirrors.

Angle of Refraction

Light is able to pass through many types of materials, including air, water, and clear crystals or plastic. However, the properties of these materials will usually cause the light to change in speed when it hits the material. As a result, the angle of the light when it exits the material will change as well. General rules of refraction:

- If light travels into a medium into which it moves slower, then the light is refracted away from the centerline.
- If light travels into a medium into which it moves faster, then the light is refracted toward the centerline.

How can we calculate how much the light is refracted? The angle at which the light is refracted is dependent on the optical density and the index of refraction of a material. The equation that is used to govern this is called Snell's Law, and is given below:

$$\frac{\sin\theta_1}{\sin\theta_2} = \frac{v1}{v2} = \frac{n2}{n1}$$

Where "v" is the velocity of light in that material and "n" is the refractive index of the material. Measurement of the speed of light in a given material is difficult, given the very high speed of light, so for most calculations, the refractive index of the material is used. A table of sample refractive indices is given below.

Material	N
Air	1
Water	1.33
Diamond	2.41
Acrylic Glass	1.49

Sample Problem – Angle of Refraction

Sunlight strikes a pane of glass at an incident angle of 60 degrees. At what angle will it leave the glass?

To solve this problem, we *could* draw a diagram showing the projected incident angle and refracted angle, but that probably isn't necessary given the simplicity of the problem.

1. We set up Snell's law to show:

$$\frac{\sin(60)}{\sin(x)} = \frac{1.49}{1}$$

Now, we can solve for x to find that the refracted angle is 35.5 degrees. If we drew a small picture of this refraction, it would look like this:

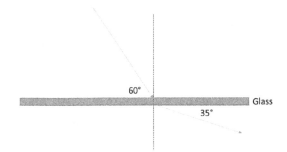

Note!

- If light travels into a medium into which it moves slower, then the light is refracted away from the centerline.
- If light travels into a medium into which it moves faster, then the light is refracted toward the centerline.

Nuclear Physics & the Model of the Atom

This part of the review will cover a basic understanding of elementary particles. This includes photons, parts of the atoms such as electrons, protons, and neutrons, and also nuclear reactions (but not in the nuclear bomb sense of nuclear reaction). The AP Physics exam does not test heavily on this subject. However, you are expected to know the basics of nuclear physics, as it relates to the properties of materials at a more macroscopic scale.

Photons

A photon is an elementary unit of light, and was originally known as a quanta, or a "packet" of energy. Interestingly, a photon is considered to have no mass when not moving. It only has mass when it has a velocity, which was a concept proposed by Albert Einstein in the 1900's. In the 1800's, when the wave-particle duality of light was discovered, there was still a problem with the math behind calculating the energy of light. It appeared that because of the dual nature of light, energy was not being conserved.

Einstein, in conjunction with scientists Maxwell and Planck, developed a theory that quantified the energy levels of light in specific and discrete steps. They believed that light could only change energy in particular discrete amounts, and could not change in energy steps incrementally. Based on this, the equation to calculate the energy of a photon was developed:

$$E = \frac{hc}{\lambda}$$

Where h is Planck's constant 6.626×10^{-34} m²*kg/s, which represents the discrete levels of energy that light can possess, c is the speed of light, 3.0×10^8 m/s, and lambda is the wavelength of the light.

Sample Problem – Nuclear Physics

What is the energy difference between violet light, which has a wavelength of 400 nm, and red light, which has a wavelength of 700 nm?

To solve this problem, we can use the relationship between energy of a photon and its wavelength.

1. First, we convert the wavelengths of the light into meters into order to obtain a correct answer. 400 nm is 4×10^{-4} meters, and 700 nanometers is 7×10^{-4} nanometers. We insert these values into the equation, and find that:

    Violet light: 4.96×10^{-22} Joules per photon

    Red light: 2.83×10^{-22} Joules per photon

    Based on this answer, we can say that violet light contains more energy than red light. Actually, red light has the lowest energy out of the visible light spectrum.

The Atom

The atom is considered the smallest fundamental building block of matter. It is made up of even smaller components, called protons, neutrons, and electrons. If you want to get detailed, each of these smaller components are made up of different types of quarks, which are extremely small subatomic particles that have been discovered relatively recently.

However, the atom is the smallest unit of matter that has unique discernable properties of each element. The properties of elements and their atoms are determined by the number of protons and electrons that exist in each atom. These details can be found in the periodic table. The group number in the periodic table represents the number of valence (outer shell) electrons found in each element, and the period number represents the total number of shells of electrons that each atom has.

For example, we will look at two common elements: oxygen and calcium.

- Oxygen is in the 2nd period, meaning it has 2 shells of electrons, and it is in group 16, which for the 2nd period means that it has 6 electrons in its valence shell.
- Calcium is in the 3rd period, meaning it has 3 shells of electrons, and it is in group 2, meaning that it has 2 electrons in its valence shell.

In the atom, the proton number determines the type of atom it is, and the electron count in its valence shell determines the reactivity of the atom.

Description of the Atom

The atom is composed of a core/nucleus of protons and neutrons that make up the majority of the mass of the atom. Electrons orbit the nucleus in orbitals. However, it is best not to think of the orbitals as circular orbitals, such as planets orbiting the sun. Instead, modern microscopy and crystallography techniques have shown that electrons form a "cloud" in an orbit-like shape around the nucleus.

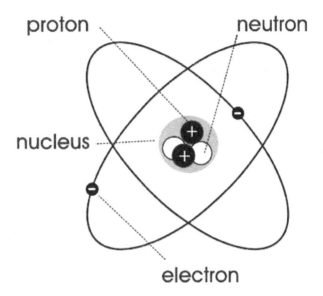

These clouds are defined by the shapes of their orbitals, and are named the "s" orbital, "p" orbital, "d" orbital, or "f" orbital. The s-orbital and p-orbital cloud shapes are shown below.

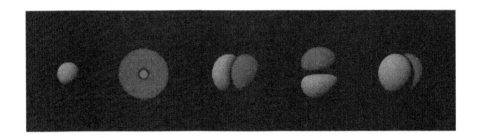

> **Note!**
> - The energy contained in light is directly proportional to its wavelength. The shorter the wavelength, the higher the energy.
> - The atom is composed of neutrons, protons, and electrons. The number of protons determines the element type, and the number of electrons in the atom determines its electronic properties.

Practice Examination

Free Response Questions: A Primer

The AP Physics B exam has a free response section that will ask 6 or 7 questions. At least one of these questions will be focused on a physics laboratory experiment. Compared to other AP exams, there are no "short-answer" or "long-answer" problems on the free response section. You will have 1 hour and 30 minutes to answer all of the questions to the best of your ability.

In the multiple choice question section, the majority of questions will not involve a significant portion of calculation. On the free response question section, nearly all the answers require some calculation, so make sure to show your work on this portion of the exam! Below is a sample free response question and the associated answer.

Sample Problem – Free Response

A cube with an unknown mass M and a side length L is submerged into a tank of water. It is attached to the bottom of the tank with a length of rope, as shown in the figure below. The tension on the string is equal to ½ of the weight of the cube. The density of water is known at 1000 kg/m^3.

- Draw a pictorial representation of the problem, and a point diagram representing the forces exerted on the cube.
- What is the density of the cube?
- If a diver swims to the bottom of the tank and cuts the rope attaching the cube to the tank, what is the initial rate of acceleration for the cube?
- As the cube rises, will the buoyant force remain constant? Why or why not?

1. In this problem, you are being tested on your knowledge of kinetics of motion, buoyant force, and rope tension. We will start by drawing the force diagram as needed in part A.

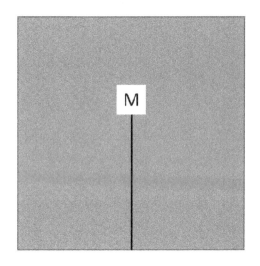
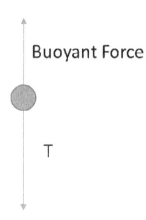

We note that there are two forces acting downward: a gravitational force m*g, and a tension force from the rope. There is a buoyant force as well, since the cube is floating.

There is also a pressure force in this scenario. However, because the pressure force is equal on all surfaces of the cube, and does not affect the buoyancy of the cube, we will ignore the pressure force for this problem.

2. Now, we can calculate the buoyant force of the cube, and from this, we can calculate the density of the cube.

From Archimedes' principle, we know that the buoyant force is equal to mass displacement of the cube. If 1 liter of water is displacement, then the buoyant force is equal to 9.8 N (the weight of the water).

The problem tells us that the tension in the rope is equal to ½ of the weight of the cube. This means that 0.5*M*g = T.

In order to get a handle on this problem, we should first think of a comparative situation. Let's imagine that this cube is floating on the surface of the water. If it has a density of 500 kg/m^3, then we know that it will float. If we want to push the cube fully under water, then we have to use a force of 500kg*9.8m/s^2, or 4900 N, in order to submerge the cube. This is the same amount of tension force that would be needed to hold the cube in the middle of the tank.

Based on this, if we choose a representative volume, we can to set up an equation to solve for the density of the cube.

$$(1 - d)g = 0.5(d \times g)$$

The left hand side of the equation represents our tension force, which is equal to the density of water minus the density of the cube, multiplied by the gravitational force. The right hand side of the equation represents the ½ of the weight of the cube. Now, if we solve for "d", then we have the density.

We know g is constant at 9.8 m/s². Now, if rearrange to solve for density, we get:

$$d = \frac{g}{1.5g}$$

Since the gravitational constants cancel each other out, then the specific density must be 0.666, which would result in 666 kg/m³.

We should check this answer quickly. If the cube has a density of 666 kg/m³ and a volume of 1 meter, then the buoyant force will be equal to 333kg*9.8, or 3263 N.

The weight of the cube is 666 kg*9.8 = 6526 N. Thus, the tension in the rope is the same as the one half of the weight of the cube, and we have solved the second part of the problem!

3. If a diver swims down to the bottom of the tank and cuts the rope, the rope tension is no longer negating the buoyant force. The new force diagram looks like:

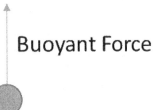

Because there is no force to counteract the buoyant force, the cube should accelerate upward. How fast? This is difficult because we don't know the mass of the cube. We do know the density however!

If the buoyant force is equal to the weight of the cube, then the cube will accelerate at 9.8 m/s².

$$F = ma = weight\ of\ cube, then\ a = g$$

If the buoyant force is equal to half the weight of the cube, then it makes sense that the cube will accelerate at half the rate. Thus, we conclude that the cube will accelerate at 4.9m/s².

Quiz 1

Multiple Choice

1. A car travels with a constant speed of 40 mi/hr. If the car travels at this speed for 15 minutes, how far does the car travel in this time?
   a) 10 miles
   b) 2.7 miles
   c) 0.375 miles
   d) 40 miles
   e) 600 miles

2. Consider a ball dropped from the top of a building. If the height of the building is 20 m and air resistance can be neglected, how long does it take the ball to reach the ground ($g \sim 10$ m/s²)?
   a) $\sqrt{2}$ s
   b) 2 s
   c) 4 s
   d) 10 s
   e) 5 s

3. The following graph shows the velocity of a moving object as a function of time during a time interval of (3)

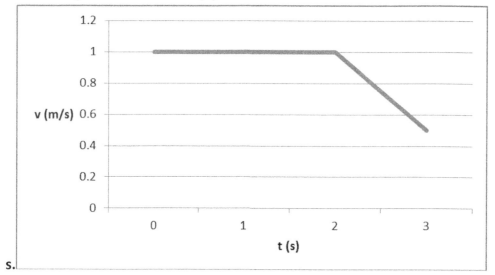

s.

Which of the following statements concerning the motion of the object is true?

a) The object remains still for $0 \leq t \leq 2$ s and then starts to move backwards.
b) The acceleration of the object remains the same for $0 \leq t \leq 3$ s.
c) The acceleration of the object is zero for the first 2 seconds and then the magnitude of the acceleration decreases.
d) The acceleration of the object is zero for the first 2 seconds and then the acceleration of the object is negative.
e) The acceleration of the object is zero and then the object moves backwards.

4. A car travels at 40 mph at 60° north of west. The component of the velocity in the west direction is:
a) 2 miles/hr
b) 20 miles/hr
c) 40 miles/hr
d) 80 miles/hr
e) 4 miles/hr

5. Two dogs are running towards each other. Dog #1 has a speed of 0.75 m/s and the dog #2 has a speed of 0.55 m/s. If we define the positive x direction to be the direction in which dog #1 runs, what is the velocity of dog #1 with respect to dog #2?
   a) 0.75 m/s
   b) 0.2 m/s
   c) -0.75 m/s
   d) -1.3 m/s
   e) 1.3 m/s

6. Which of the following statements concerning projectile motion is *not* true? Note that any effects due to air resistance are neglected.
   a) The horizontal velocity of an object launched upwards at an angle to the horizontal is constant.
   b) The acceleration is positive while the object moves upwards and negative while the object moves downward.
   c) The acceleration of the object is always negative.
   d) The vertical component of the velocity of an object launched upwards at an angle to the horizontal is zero at the highest point of the object's motion.
   e) The acceleration of the object in the horizontal direction is zero.

7. Consider an object acted on by only 2 forces as shown. If the magnitudes of F_1 and F_2 are equal, which of the following statements *must* be true?

   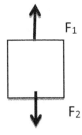

   a) The velocity of the object must be zero.
   b) The velocity of the object must be constant.
   c) The velocity of the object must be increasing.
   d) The velocity of the object must be decreasing.
   e) The object must remain stationary.

8. A child pulls a wagon along the ground with a force F_1 of magnitude 30 N. The wagon has a mass of 0.5 kg and moves with constant acceleration of 1 m/s² in the +x direction. What is the magnitude of the force F_2 (in N) acting in the −x direction?

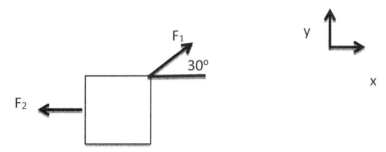

a) 30cos(30°) +0.5
b) 30cos(30°) − 0.5
c) 0.5 − 30sin(30°)
d) 30cos(30°)
e) 30sin(30°)

9. An object is pulled across a rough surface by a force of 20 N. If the object moves with a constant velocity and the surface has a coefficient of kinetic friction $\mu_k = 0.2$, what is the magnitude of the normal force acting on the object?
a) 100 N
b) 10 N
c) 1 N
d) 20 N
e) 4 N

10. A system of 3 blocks rest on a frictionless surface as shown below. A force of magnitude f is applied to block 3 as shown which causes the system of blocks to accelerate with an acceleration a. If $m_1 = 3m_3$, the magnitude of the force acting on block 1 is:

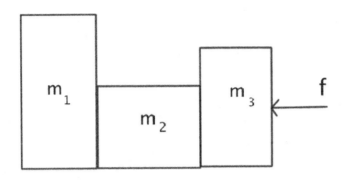

a) f
b) $f/3$
c) $3f$
d) f^3
e) m_1/a

For questions 11-12, refer to the following scenario: An object of mass *m* is pushed up a frictionless inclined plane by a force of magnitude f_1 as shown.

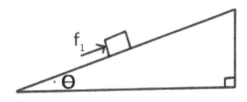

11. Which of the following expressions gives the net force acting up the slope?
    a) $f_1 - mg$
    b) $f_1 + mg$
    c) $f_1 - mg\cos\theta$
    d) $f_1 - mg\sin\theta$
    e) $f_1 + mg\sin\theta$

12. Suppose that instead of a force f_1 pushing the object up the slope, the object, instead, is attached to a rope which passes over a pulley and is attached to a mass m at its other end. *T* denotes the tension in the rope. The object now remains stationary. Which of the equations below is *not* valid for this situation?

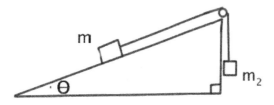

a) $T = mg\sin\theta$
b) $T = m_2 g$
c) $m_2 = m\sin\theta$
d) $T = (m_1 + m_2)a$
e) $N = mg\cos\theta$

13. An object of mass *m* undergoes uniform circular motion with a speed *v* at a distance *r* from the center of the circle. Which of the following will result in an increase in the centripetal force?
   a) An increase in *r* while maintaining the same speed.
   b) A decrease in the speed of the mass while maintaining the same distance *r*.
   c) An increase in the speed of the mass by a factor of 2 and a decrease in *r* by a factor of 4.
   d) A decrease in *r* while maintaining the same speed.
   e) An increase in the speed of the mass by a factor of 2 and an increase in *r* by a factor of 4.

14. A ball of mass m is attached to a rope and then swung in a vertical circle. Which of the following equations is valid at the bottom of the circle?

a) $m\dfrac{v^2}{r} = T - mg$

b) $m\dfrac{v^2}{r} = T + mg$

c) $m\dfrac{v^2}{r} = T$

d) $m\dfrac{v^2}{r} = mg$

e) $T = mg$

15. An object rotates about point B as shown. A force of magnitude 10 N is applied at point A. What is the torque produced by the force?

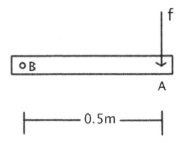

a) 5 Nm
b) −5Nm
c) 20 Nm
d) −20 Nm
e) 10 Nm

For questions 16-17, refer to the following scenario: A child pushes a 0.5 kg truck across the floor. The truck is initially at rest and reaches a final velocity of 1 m/s.

16. How much work does the child do?

a) 0.25 J
b) 2.5 J
c) 25 J
d) 50 J
e) 5 J

17. If it takes 5s for the truck to reach this speed, how much power did the child deliver to the truck?

    a) 5 W
    b) 0.5 W
    c) 2.5 W
    d) 0.05 W
    e) 0.25 W

18. A skier starts from rest at the top of a ski hill. If the height of the ski hill is 20 m and any effects due to air resistance or friction are ignored, what is the speed of the skier at the bottom of the hill?

    a) 200 m/s
    b) 20 m/s
    c) 14 m/s
    d) 400 m/s
    e) 4 m/s

19. What is the momentum of an object of mass $m=15$ kg that moves with a velocity of magnitude 10 m/s in the −y direction?

    a) 15 kg m/s
    b) 150 kg m/s
    c) -15 kg m/s
    d) -150 kg m/s
    e) 1.5 kg m/s

20. Consider a system of 2 objects. If both internal and external forces act on the system and if the net external force on the system is zero, which of the following statements *must* be true?

    a) The internal forces cause a change in the net momentum of the system.
    b) The momentum of both objects remains unchanged.
    c) The momentum of one object changes while the other stays the same.
    d) Both objects must be stationary.
    e) The change in the net momentum of the system is zero.

21. Two objects travel in the +x direction with different speeds. Object 1 moves at 5 m/s and has a mass of 1 kg. Object 2 moves at a slower speed of 2 m/s and has a mass of 0.5 kg. When Object 1 collides with object 2, they stick together. What is the final velocity of the two objects?

    a) 2.5 m/s
    b) 4 m/s
    c) -4 m/s
    d) -2.5 m/s
    e) 7 m/s

22. Which of the following statement is true concerning an object undergoing simple harmonic motion, such as a mass on the end of a spring?

    a) The kinetic energy of the system is 0 at the midpoint of the motion.
    b) The potential energy of the system is always 0.
    c) The speed of the mass is constant during the motion.
    d) The kinetic energy of the system is maximum at the midpoint of the motion.
    e) The potential energy of the system is maximum at the midpoint of the motion.

23. Consider a mass on the end of a spring. The spring has a spring constant of 80 N/m. The spring is compressed a distance 10 cm and then let go and the system undergoes Simple Harmonic Motion. What is the total energy of the system?

    a) 0.4 J
    b) 4 J
    c) 0.8 J
    d) 10 J
    e) 4,000 J

24. Consider two objects a distance *r* apart. According to Newton's Law of Universal Gravitation, what happens to the force between the two objects if the distance r is increased by a factor of 4?

    a) the force decreases by a factor of 4.
    b) the force increases by a factor of 4.
    c) the force increases by a factor of 2.
    d) the force decreases by a factor of 16.
    e) the force decreases by a factor of 32.

25. Consider a container of water and the points A, B, C, and D as shown. Which statement is true?

    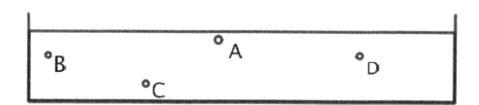

    a) The pressure at B is less than the pressure at A
    b) The pressure at A is equal to the pressure at C
    c) The pressure at B and D are equal
    d) The pressure at D is greater than the pressure at C
    e) The pressure at B is greater than pressure at C

For questions 26 and 27, consider an object with a volume of 10 m² and density ρ_o floating in water (ρ_w).

26. If the object is floating in fresh water and the ratio $\frac{\rho_o}{\rho_w} = 0.6$, what volume of water is displaced?
    a) 10 m³
    b) 6 m³
    c) 60 m³
    d) 0.06 m³
    e) 16 m³

27. If the object is floating in saltwater instead of freshwater, how does the volume of water displaced by the object compare to the volume of water that is displaced when the object floats in fresh water?

    a) The volume of water displaced remains the same.
    b) The volume of water displaced increases.
    c) The volume of water displaced decreases.
    d) The volume of water displaced is equal to the volume of the object.
    e) The volume of water displaced first increases then decreases.

For questions 28 and 29, consider an incompressible fluid (density remains constant) flowing through a tube. The fluid flows through a section of the tube with a radius r_1 and then into a section of the tube with a radius r_2.

28. If $r_2 > r_1$, the effect of the larger radius on the flow is to:

    a) Increase the volume of fluid that flows past a given point.
    b) Increase the speed of the fluid.
    c) Decrease the volume of fluid that flows past a given point.
    d) Decrease the speed of the fluid.
    e) Increase the speed of the fluid by r_2/r_1.

29. If the speed through the section of tube with radius r_1 is 4 m/s, what value of r_2 is required so that the speed through that section of the tube is 1 m/s?

    a) $4r_1$
    b) $r_1/4$
    c) $2r_1$
    d) $r_1/2$
    e) $8r_1$

30. The volume of an object is 50 cm³ at 274 K. The object is made of a material with a coefficient of thermal expansion $\alpha = 1 \times 10^{-3}$. At what temperature would the volume of the object equal approximately 50.05 cm³?

    a) 307 K
    b) 30 K
    c) 3 K
    d) 3000 K
    e) 240 K

31. Which of the following statements is *not* true concerning ideal gases?
    a) The equation $PV = nRT$ applies.
    b) In an ideal gas, the molecules do not interact.
    c) The behavior of an ideal gas is a good approximation for the behavior of most real gases.
    d) The pressure of an ideal gas increases as the temperature is increased.
    e) An ideal gas will become solid at low temperatures.

32. Consider a system in which both the heat gained or lost by the system and the work done by or on the system can change. The system can either do work on the environment or have work done on it and either loses or gains heat. Which of the following situations *must* result in an increase in the internal energy of the system?
    a) Work is done on the environment and the system gains heat.
    b) Work is done by the environment on the system and the system gains heat.
    c) Work is done on the environment and the system loses heat
    d) Work is done by the environment on the system and the system loses heat
    e) Work is done by the environment on the system but it is unknown whether the system gains or loses heat

33. The definition of an adiabatic process is a process in which
    a) the pressure of the system is constant.
    b) the volume of the system is constant.
    c) no heat flows into or out of the system.
    d) the temperature of the system is constant.
    e) the product of the pressure and the volume is a constant.

34. Consider the following graph showing the pressure and volume of an ideal gas during a process. What is the work done by the gas during the process?

    a) 25 J
    b) 50 J
    c) 100 J
    d) 0.5 J
    e) 100 J

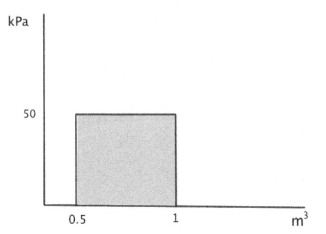

35. A heat engine operates with a high temperature of 500K and a low temperature of 200K. What is the maximum efficiency of the heat engine?
    a) 6
    b) 0.6
    c) 2
    d) 60
    e) 0.06

36. Two particles are separated by a distance of 0.5 m. If one particle has a charge of 1 µC and the other particle has a charge of -2 µC, what is the magnitude and direction of the force between the particles? ($k = 9 \times 10^9$ Nm²/C²)
    a) 7.2 N
    b) 0.72 N
    c) 0.072 N
    d) 72 N
    e) 720 N

37. A particle is placed in a uniform electric field of magnitude 1×10^4 N/C. If the particle experiences a force of magnitude 0.1 N, what is the magnitude of the charge on the particle?
    a) 10 µC
    b) 1 µC
    c) 0.1 µC
    d) 100 µC
    e) 1 C

38. Which of the following statements is true concerning a negative point charge and the electric field it produces?
    a) The magnitude of the electric field depends on the magnitude of the point charge and the field is directed radially outwards.
    b) The magnitude of the electric field does not depend on the magnitude of the point charge and the field is directed radially outwards.
    c) An electric field is not produced by a negative point charge.
    d) The magnitude of the electric field depends on the magnitude of the point charge and increases as the distance from the point charge increases.
    e) The magnitude of the electric field depends on the magnitude of the point charge and the field is directed radially inwards.

For questions 39-40, consider the following scenario: A uniform electric field has a magnitude of 1×10^4 N/C and is directed to the right. A test charge Q of 1×10^{-6} C is moved a distance 10 cm to the left.

39. What is the work done by the electric field?
    a) 1 J
    b) 0.1 J
    c) 0.001 J
    d) -0.001 J
    e) -1 J

40. What is the change in the electric potential?
    a) 1×10^3 V
    b) -1×10^3 V
    c) 1×10^2 V
    d) -1×10^2 V
    e) 10 V

41. Which of the following statements concerning ideal conductors is *not* true?
    a) The potential at every position on or within an ideal conductor is the same.
    b) The potential at every position within an ideal conductor must be zero.
    c) Any excess charge that an ideal conductor contains resides on its surface when in equilibrium.
    d) The electric field within an ideal conductor is zero.
    e) Charges are free to move in a conductor.

42. A capacitor consists of two parallel plates of area A separated by a distance d. If the distance d is set at 1×10^{-4} m and a capacitance of 1×10^{-11} F is needed, what is the area required?
    a) 1×10^{-2} m²
    b) 1 m²
    c) 1×10^{2} m²
    d) 1×10^{-4} m²
    e) 1×10^{3} m²

For questions 43-45, consider the following: A series circuit is shown below. A voltage of 20 V is applied to the circuit. The values of the resistors are:

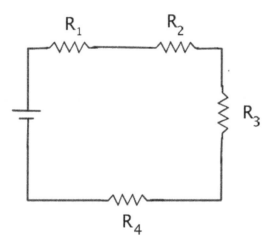

$R_1 = 100\ \Omega, R_2 = 250\ \Omega, R_3 = 150\ \Omega, R_4 = 100\ \Omega.$.

43. What is the current flowing through the circuit?

    a) 30 A
    b) 1/30 A
    c) 600 A
    d) 1/3 A
    e) 300 A

44. What is the voltage drop across R_3?

    a) 50 V
    b) 20 V
    c) 1.5 V
    d) 2 V
    e) 5 V

45. The circuit is modified as shown. R_5= 500Ω. How is the current through R_1 affected?

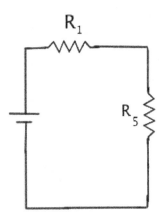

    a) The current through R_1 increases.
    b) The current through R_1 increases.
    c) The current through R_1 is not changed.
    d) The current through R_1 is reduced to 0.
    e) The current through R_1 is doubled.

46. Consider the circuit shown. Which of the following statements is true concerning the placement and design of ohmmeters and ammeters to achieve correct readings?

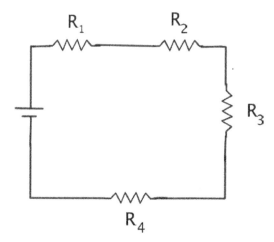

a) An ideal ammeter is placed in parallel with R_1 and has an infinite resistance.
b) An ideal ammeter is placed in series with R_1 and has an infinite resistance.
c) An ideal voltmeter is placed in series with R_1 and has an infinite resistance.
d) An ideal voltmeter is placed in parallel with R_1 and has an infinite resistance.
e) An ideal ammeter is placed in parallel with R_1 and has zero resistance.

47. The following circuit contains two capacitors. If $C_1 = 2 \times 10^{-10}$ F and $C_2 = 3 \times 10^{-10}$ F, what is the equivalent resistance?

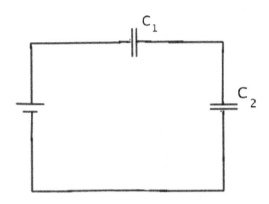

a) 5×10^{-10} F
b) 1×10^{-10} F
c) 0.8×10^{10} F
d) 5×10^{-5} F
e) 0.8×10^{20} F

48. A charge Q is placed in a magnetic field of magnitude B, directed in the +y direction. Under which of the following conditions will the charge Q experience a force due to the magnetic field?
   a) The charge Q is stationary.
   b) The charge Q moves in the +y direction.
   c) The charge Q moves in the –y direction.
   d) The charge Q moves in the +x direction.
   e) The charge Q = 0, that is, a neutral particle and moves in the +x direction.

49. A wire of length 2 m carries a current of 5A in the –x direction. A magnetic field of 1T points in the +y direction. What is the magnitude and direction of the force on the wire due to the magnetic field?
   a) 10 N in the +x direction.
   b) 10 N in the –y direction.
   c) 10 N directed out of the page.
   d) 10 N directed into the page.
   e) 10 N in the +y direction.

50. Two long current-carrying wires are placed parallel to each other a distance *d* apart. Which of the following situations would result in the largest magnetic field between the wires?
    a) Two identical wires varying the same current of magnitude I in the same direction.
    b) Two identical wires, one carrying a current of magnitude I and the other a current of magnitude 2I, both in the same direction.
    c) Two identical wires, one carrying a current of magnitude I and the other a current of magnitude 2I, in opposite directions.
    d) Two identical wires, one carrying a current of magnitude I and the other a current of magnitude I/2, in opposite directions.
    e) Two identical wires, one carrying a current of magnitude I and the other a current of magnitude I/2, in the same direction.

51. A coil of wire is placed in a magnetic field of 2T directed in the -x direction. If the coil has an area of 0.1 m² and is placed such that it makes an angle of 60° with the magnetic field, what is the flux through the coil?
    a) 0.1 Wb
    b) 1 Wb
    c) 10 Wb
    d) 0.01 Wb
    e) 100 Wb

52. A wire coil is placed in a magnetic field B and connected to a circuit. Which of the following will result in a current flowing through the circuit?
    a) The magnetic flux through the coil is constant.
    b) The magnetic field is uniform and at a 90° angle to the normal of the coil.
    c) The coil is placed such that the magnetic field is parallel to the normal of the coil.
    d) The magnitude of the magnetic field changes with time.
    e) The magnetic flux through the coil is zero.

53. A closed loop of wire is held horizontally in a magnetic field directed upwards (perpendicular to the horizontal plane) such that the magnetic flux through the coil decreases. Which of the following best describes the effect on the coil?
    a) There is no effect on the coil.
    b) A current flows clockwise within the coil.
    c) A current flows counterclockwise within the coil.
    d) No current flows within the coil.
    e) The coil will expand.

54. Which of the following is an example of a longitudinal wave?
    a) A wave on a string
    b) A light wave
    c) A sound wave
    d) Standing waves on a string
    e) Water waves

55. Consider a wave propagating through a medium with a speed v such that it takes a time t_1 for the wave to travel one wavelength. If the source of the wave is altered such that a new wave is created with a longer wavelength than the original one, which of the following statements is true when comparing the new wave to the original wave? Note that the speeds of both waves are the same.
    a) The frequency of the new wave is greater than the frequency of the original wave
    b) The frequency of the new wave is the same as the frequency of the original wave.
    c) The period of the new wave is less than the frequency of the original wave.
    d) The frequency of the new wave is less than the frequency of the original wave
    e) The period of the new wave is the same as the period of the original wave.

56. A string of length 1 m is attached at both ends to solid surfaces. A standing wave is produced on the string such that the string vibrates in its first harmonic. What is the wavelength of the wave?
    a) 2 m
    b) 1 m
    c) 0.5 m
    d) 4 m
    e) 0.25 m

57. How does the first harmonic for a wave on a string of length L compare to the first harmonic for a sound wave in a tube of length L open at one end and closed at the other?
    a) The wavelengths of the first harmonics are the same.
    b) The wavelength of the first harmonic for a tube open at one end is twice the wavelength of the first harmonic for a wave on a string.
    c) The wavelength of the first harmonic for a tube open at one end is less than the wavelength of the first harmonic for a wave on a string.
    d) The wavelength of the first harmonic for a tube open at one end is four times as long the wavelength of the first harmonic for a wave on a string.
    e) The wavelength of the first harmonic for a tube open at one end is half the wavelength of the first harmonic for a wave on a string.

58. The following is a displacement-time graph for a wave. The amplitude of the wave is:

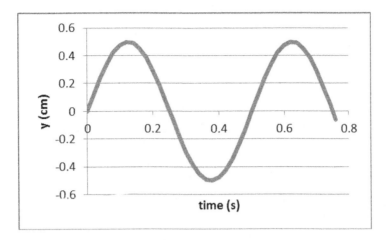

a) 1 cm
b) 0.5 cm
c) 0.4 s
d) 0.8 s
e) 0 cm

59. Two coherent waves each with amplitude A and wavelength λ are produced by different sources. The waves are in phase with each other and arrive at a point P. If one wave travels a distance of $\lambda/2$ from its source to the point P and the other wave travels a distance of $5\lambda/2$, what is the amplitude of the wave at point P?
a) 0
b) 2A
c) A
d) A/2
e) 4A

For questions 60 and 61, consider a ray of light directed at a reflective surface as shown.

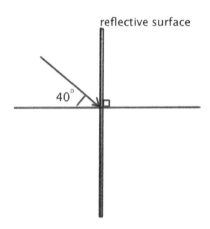

60. What angle does the reflected ray make with respect to the normal?

    a. 50°
    b. 40°
    c. 0°
    d. 90°
    e. 20°

61. If the reflective surface is rough, not smooth, what is the effect on the reflected ray?

    a) The reflected ray obeys the Law of Reflection regardless of the reflective surface.
    b) The ray will not be reflected at any angle.
    c) The ray is reflected at an angle most likely different from the incident angle depending on where it strikes the surface.
    d) The angle of reflection will equal the angle of incidence.
    e) The ray will always be reflected at right angles to the incident ray.

For questions 62-63, consider the following: Light travels at a speed v_1 through a medium with a refractive index n_1. The light then enters a second medium with a refractive index n_2.

62. If the ratio $\frac{n_1}{n_2} > 1$, which of the following statements is true?

    a) The light will not enter the second medium
    b) The speed of light in the second medium will be less than in the first medium.
    c) The speed of light in the second medium will be greater than in the first medium.
    d) The light will be have the same speed in both mediums.
    e) The light will be totally reflected back into the first medium.

63. Under which condition will the light not be deflected from its path as it travels from the first medium to the second medium?

    a) The angle at which light enters the second medium with respect to the normal is less than 45°.
    b) The angle at which light enters the second medium with respect to the normal is greater than 45°.
    c) The angle at which light enters the second medium with respect to the normal is equal to 45°.
    d) The angle at which light enters the second medium with respect to the normal is equal to 0°.
    e) The light will always be deflected as it enters the second medium.

64. Consider an image formed by a plane mirror. The properties of the image are such that:

    a) The image is inverted and real.
    b) The image is inverted and virtual.
    c) The image is upright and real.
    d) The image is upright and virtual.
    e) The image is upright, virtual, and bigger than the object.

65. Which of the following is the most likely value for the energy of a photon from the visible part of the electromagnetic spectrum?

   a) 1000 eV
   b) 100 eV
   c) 1 eV
   d) 0.01 eV
   e) 0.001 eV

66. Which of the following statements correctly describes a photon?

   a) A photon has mass when at rest and travels at the speed of light.
   b) A photon has mass when at rest and travels slower than the speed of light.
   c) A photon has mass when at rest but no momentum.
   d) A photon has zero mass when at rest and travels at the speed of light.
   e) A photon has zero mass when at rest and zero momentum when traveling.

67. An electron has a momentum of 1×10^{-27} kg m/s. What is best estimate of the de Broglie wavelength? (Note: take Plank's constant to be approximately 7×10^{-34} Js)

   a) 7×10^{-7} m
   b) 7 m
   c) 7×10^{7} m
   d) 7×10^{-3} m
   e) 0.7 m

68. The nucleus of sodium is represented as $^{22}_{11}Na$. Which of the following correctly describes the nucleus?

    a) The number of nucleons is 22 and the protons plus neutrons is 11.
    b) The number of nucleons is 22 and the number of protons is 11.
    c) The number of protons is 11 and the number of neutrons is 22.
    d) The number of nucleons is 11 and the number of neutrons is 22.
    e) The number of neutrons is 11 and the number of protons is 22.

69. An unstable nucleus experiences alpha decay. Which of the following occurs during or is the result of the process?

    a) A photon is emitted.
    b) A positron is emitted.
    c) The atomic number decreases by 1.
    d) The atomic number decreases by 2.
    e) The mass number decreases by 1.

70. The length of a slab of material is measured using a ruler on which the smallest division is 0.5 mm. What is the best estimate for the uncertainty involved in a measurement done with this ruler?

    a) 0.2 mm
    b) 0.5 mm
    c) 0.4 mm
    d) 1 mm
    e) 0.05 mm

Free Response Questions

1. An object is launched with a speed of 5 m/s at an angle $20°$ to the horizontal as shown.

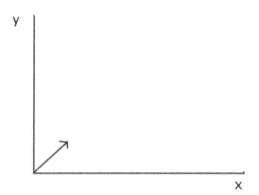

   i. Calculate the horizontal distance traveled by the object.

   ii. Determine the time it takes for the object to reach its highest point.

   iii.

         a. What is the horizontal speed of the ball when it lands?

         b. Explain why this is so.

2. An object of mass 1 kg rests on a rough surface ($\mu_s = 0.5$). A man applies a force of magnitude f to the box directed to the right in the picture below. The box does not move under the application of the force.

   a. Draw the free-body diagram for the box, indicating all forces acting on the object.

   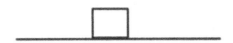

   b. Explain why the object is not moving when the force of magnitude f is applied to it.
   c. What force does the man need to apply for the object to begin to move?

   d. Once the man has started the object moving, the man attempts to push the object up a slope. The slope makes an angle θ= 45° and the surface roughness is the same as the roughness of the level surface. The man tires out partway up the slope and stops pushing. Does the object stay still or slide back down the slope? Explain.

   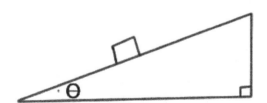

3. A mass m = 0.2 kg is attached to the end of a spring. The other end of the spring is attached to a solid surface and the spring hangs vertically such that the mass is free to oscillate. The force constant of the spring is 10 N/m.
    a. By how much is the spring stretched from its unstretched length when in equilibrium?
    b. Suppose the spring is now disturbed from its equilibrium position so that the mass oscillates in Simple Harmonic Motion. Calculate the period of the oscillation?
    c. Below is a plot of the oscillation about the equilibrium position. Sketch the plot if the force constant is increased by a factor of 4.

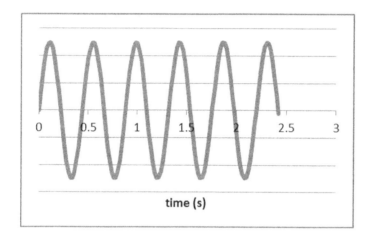

4. An unknown liquid is investigated to determine its properties. A container holds a volume of the liquid and is open at the top. A sensor is placed at a depth of 0.1 m and provides information about the pressure. The pressure at 0.1 m is determined to be 1.03×10^5 Pa. What will be the pressure a further 1 m below this?

5. Consider 3 charges placed along a straight line. A positive charge q_1 of 2μC is placed at position x_1= 0.5 m from the origin and a second positive charge q_2 of 3 μC is placed at position x_2.= 1 m from the origin as shown. A third charge of -1 μC is placed in between the two positive charges. At what x position should this negative charge be placed such that it is in equilibrium?

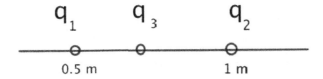

6. A Silicon nucleus undergoes beta decay.
    a. Complete the following nuclear reaction.

    $$^{31}_{14}Si \rightarrow {}^{\square}_{\square}P + e^-$$

    b. What is the energy released by the silicon nucleus during this reaction (mass of the silicon nucleus = 30.975 u, mass of the phosphorus nucleus = 30.974u)?

7. A person singing produces sound with a power of 0.09 W.
    a. An observer sits a distance of 5 m away from the singer. What is the intensity of the sound heard by the observer?
    b. The lowest intensities humans can hear are of the order 10^{-12} W/m². Calculate an estimate of how far an observer would have to move away from the singer before the observer could not detect any sound from the singer. Assume no other sounds interfere with the sound produced by the singer or heard by the observer.

8. Describe and explain the Doppler effect.

Answer Key 1

Multiple Choice

1. **Answer: a**

   Speed = distance divided by time. Here, we know the speed and time and need to calculate the distance. Distance = speed × time. The time is given to us in minutes, so we need to convert to hours. 15 minutes = 0.25 hours. Therefore, distance = 40mi/hr × 0.25 hrs = 10 miles.

   Options b-e do not provide the correct speed when substituted into the formula.

2. **Answer: b**

   We can use the equation $y = y_o + v_o t + \frac{1}{2}at^2$ because the ball is in free fall and the acceleration is due to gravity (constant acceleration). We know $y - y_o = 20$ m and $a = g \sim 10$ m/s^2. Substituting these values into the equation and solving for t, we find $t = 2$ s.

   Options a, c-e do not provide the correct height when substituted into $y = y_o + v_o t + \frac{1}{2}at^2$.

3. **Answer: d**

   Observing the graph, we see that the speed of the object is 1 m/s for the first 2 seconds. The speed then decreases from 1 m/s to 0.5 m/s over the following second. The decreasing speed during the last second means that the object is decelerating or, in other words, accelerating in the direction opposite to the motion. This can also be seen by observing that the slope of the line during this time is negative. Option d is the only option that describes this motion.

   Option a says that the object is stationary during the first 2 seconds. This is not the case as the speed is 1 m/s during this time.

   Option b says that the object has the same acceleration during the 3 seconds shown in the plot. This is not the case as the acceleration is zero during the first 2 seconds.

   Option c says that the magnitude of the acceleration decreases after 2 seconds. This is not the case because initially the acceleration is zero and then the speed changes, resulting in acceleration. Regardless of whether the acceleration is positive or negative, the magnitude is greater than zero.

   Option e says the object moves backwards after 2 seconds. Assuming the object is moving in the positive direction, the acceleration is negative, not the position.

4. **Answer: b**

The velocity is 40 miles/hr 60° north of west. We find the component of the velocity in along the axis west by 40 × cos 60°. cos 60° = ½. Therefore, the component of the velocity in the west direction is 20 mi/hr.

Options a and e are both an order of magnitude too small. Options c and d cannot be correct because they both have values larger than 40 mi/hr.

5. **Answer: e**

To find the velocity of dog 1 with respect to dog 2, v_{12}, we can use the equation $v_{12}=v_{1g}+v_{g2}$. $v_{g2} = -v_{2g}$, so we can write $v_{12}=v_{1g}-v_{2g}$. The positive x direction is taken to be the direction in which dog #1 runs. Therefore, v_{1g}=0.75m/s and v_{2g}=-0.55 m/s. The velocity of dog 1 with respect to dog 2, then, is v_{12}= 0.75 m/s +0.55 m/s = 1.3 m/s.

Options a-d find other velocities: option a gives the speed of dog 1 relative to the ground, not relative to dog 2; option b gives the difference between the two speeds; option c gives the speed of the ground relative to dog 2, option d gives the speed of dog 2 relative to dog 1.

6. **Answer: b**

For projectile motion, the acceleration felt by the object is the force due to gravity. It is always negative, independent of the objects motion.

Options a and c-e are all true for an object launched at an angle to the horizontal which then undergoes projectile motion.

7. **Answer: b**

   Newton's first law tells us that if there is not net force acting on an object, the velocity of the object will not change. If the object is stationary, a net force is required for it to move; if an object moves with constant velocity, a net force is required to change that velocity. The object in this case is acted on by only 2 forces. Both forces have equal magnitude but act in opposite directions. Therefore, the net force on the object is zero. We are not told if the object is initially at rest or is moving. All we can say for sure is that the net force on the object is zero and so the velocity of the object must remain unchanged.

   Options a and e could potentially be true. However, we are not told if the object is initially at rest and therefore, cannot say whether it stays at rest.

   Options c and d are incorrect because the velocity must remain constant if no net force acts on the object.

8. **Answer: b**

   The wagon moves with an acceleration of 0.5 m/s² in the +x direction. Therefore, the net force on the wagon acts in the x direction and has a magnitude equal to *ma*. The force **F₂** acts in the −x direction and the x component of **F₁** acts in the +x direction. $F_{1x} = F_1 \cos(30°)$.

   We can express all this information as:
   $$ma = F_1 \cos(30°) - F_2$$
   $$(0.5 \, kg)(1 \, m/s^2) = (30 \, N)\cos(30°) - F_2.$$

   Rearranging, we get
   $$F_2 = (30\cos 30° - 0.5) \, N.$$

9. **Answer: a**

    The object moves with constant velocity. This tells us that the acceleration of the object is equal to 0 and that the net force on the object in the direction of movement is zero. Since the object is pulled with a force of 20 N, the magnitude of the frictional force opposing the motion must also be 20 N. $f_k = \mu_k N$, so $N = f_k/\mu_k$. Substituting in the values, we find that $f_k = 100$ N.

    Options b-e are all one or two orders of magnitude too small.
    Options a and c do not have the signs correct.

10. **Answer: c**

    Block 3 moves with an acceleration a due to a force f. We know $f = m_3 a$. Solving for a, we find $a = f/m_3$. All 3 of the blocks move with the same acceleration, therefore we can write $f_1 = m_1 a$. In the question we are told that $m_1 = 3m_3$. Putting this into our equation for f_1, we find $f_1 = m_1 a = 3m_3 \frac{f}{m_3}$. Finally, we find $f_1 = 3f$.

    Options d and e assume the net force is zero. Option e also calculates the x component of **F₁** wrong.

11. **Answer: d**

    The force **f₁** acts up the slope. There is also a component of the weight which acts down the slope. This component can be found to be Wsinθ. Since we are looking for the net force acting up the slope, we can define up the slope to be the positive direction and down the slope to be the negative direction. The magnitude of the net force up the slope then is f₁ – mgsinθ.

12. **Answer: d**

    Options a and b follow directly from applying Newton's laws to each mass. Option c is derived from combining options a and b. Option e is also a result of applying newton's law to the mass m on the slope in the direction perpendicular to the slope. Option d, however, is not valid for the situation and does not agree with Newton's laws.

13. Answer: d

The magnitude of the centripetal force $f_{cp} = mv^2/r$. If r is decreased and v is held constant, then f_{cp} will increase.

Options a-b result in an increase in f_{cp} and options c and e produce no change in f_{cp}.

14. Answer: a

At the bottom of the circle, the weight of the ball acts vertically downward and the tension in the string acts upward.

The net force is towards the center of the circle, which, in this case, is directed upwards. Therefore, we have:

$$m\frac{v^2}{r} = T - mg$$

Options b-e are all incorrect applications of Newton's Law. Option b involves the wrong sign for mg; option c and d both miss one force; option e neglects the fact that the ball is accelerating towards the center of the circle.

15. Answer: b

The torque is τ = rF when r is the perpendicular distance from the point where the force is applied to the axis of rotation. We also have to consider whether the torque is positive or negative. It is positive if the force causes the object to rotate counterclockwise and negative if the force causes a clockwise rotation. In this case, the force would result in a clockwise rotation and so the torque is negative. τ = -(0.5 m)(10 N) = -5 Nm.

Option a gives a positive torque, which, as discussed above, is not the case. Options c-e all have incorrect values for the torque.

16. Answer: a

W = ΔK, where K is the kinetic energy. We know that the truck was initially at rest, so the initial kinetic energy of the truck is zero. The final kinetic energy is given by $\frac{1}{2}mv^2$.

W = $\frac{1}{2}(0.5)(1^2)$ = 0.25 J.

Options b-e are all at least one order of magnitude too big.

17. Answer: d

P = W/t = 0.25J/5s = 0.05 W.

Options a-c and option e are all at least one order of magnitude too big.

18. Answer: b

In the absence of friction and air resistance, energy is conserved. At the top of the hill, the skier is at rest, so all the energy is in the form of potential energy *mgh*. At the bottom of the hill, all the energy is kinetic energy $\frac{1}{2}mv^2$. This leads to:

$$mgh = \frac{1}{2}mv^2.$$

The m's cancel, giving 2gh = v^2. So, $v = \sqrt{2gh}$. Substituting the values for *g* and *h* into the equation, we get $= \sqrt{2 \times 10 \times 20} = \sqrt{400}$ = 20. The speed of the skier at the bottom of the hill is 20 m/s.

Options a and c-e do not conserve energy. This can be seen by substituting the values into the above equation.

19. Answer: d

Momentum is a vector quantity with a magnitude given by *mv*. In this case, the velocity of the object is negative, resulting in a momentum in the –y direction. The momentum = -(15)(10) = -150 kg m/s.

Options a, c, and e are not the correct order of magnitude. Option b is correct for the magnitude of the momentum but not the direction.

20. Answer: e

If the net external force on the system is zero, momentum is conserved.

Options a and c are not true. The change in the net momentum must be zero. Option b may or may not happen. However, we cannot say for sure. All we know is that the net momentum remains unchanged. Option d is not true. The objects are not required to be stationary.

21. Answer: b

The momentum of the system before the collision is:

(l kg)(5 m/s) + (0.5 kg)(2 m/s).

Note that both velocities are positive because both objects are moving in the +x direction. After the collision, the objects stick together and move with a speed v. The momentum after the collision is (1.5 kg) v. Momentum is conserved and so we can write

(l kg)(5 m/s) + (0.5 kg)(2 m/s) = (1.5 kg) v.

6 kg m/s = 1.5 kg v

Solving for v, we find v = 4 m/s.

Options a and c-e do not conserve momentum.

22. Answer: d

The motion of a mass on the end of a spring undergoing simple harmonic motion has a maximum speed at the midpoint of the motion. Therefore, the kinetic energy is maximum at the midpoint.

Options a-c and e do not hold true for SHM.

23. Answer: a

Energy is conserved during SHM. Since we know the maximum distance the spring is compressed, we can calculate the energy stored in the system when the spring is compressed. (½)80(0.1)² = 0.4 J

Options b, d, and e are at least one order of magnitude too large. Option c is not correct as the factor of ½ has been omitted.

24. Answer: d

Newton's Law of Universal Gravitation states that the magnitude of the force of attraction between two objects F:

$$F \propto 1/r^2.$$

If *r* becomes 4*r*, then the force decreases by a factor of 16.

Options a-c and e do not agree with $F \propto 1/r^2$. Option a implies that the force is inversely proportional to *r*. Option b is not correct as the force is not proportional to *r*. Option c is not correct as this implies the force decreases by a factor of $\sqrt{r}$. Option e is not correct as the force is not proportional to $\sqrt{r}$.

25. Answer: c

The pressure in a liquid varies with depth according to P= P$_{at}$ +ρgh. Points B and D are at the same depth, so the pressure is the same at both points.

Options a, b, d, and e do not agree with this principle.

26. Answer: b

Since the object is floating, we know that the weight of water that is displaced must be equal to the weight of the object. $\rho_w V_w g = \rho_o V_o g$. The *g*'s cancel and we can rearrange the equation to give $V_w = \frac{\rho_o}{\rho_w} V_o$. Now, we can substitute in the values.

$$V_w = 0.6 \times 10 = 6 \, m^3$$

Option a is the volume of the object, not the volume of water displaced. Option c is an order of magnitude too large. Option d is an order of magnitude too small. Option e is not the correct order of magnitude.

27. Answer: c

The density of saltwater is greater than the density of freshwater and so the ratio $\frac{\rho_o}{\rho_w}$ is less than the same ratio for freshwater. Since the ratio is less, $V_w = \frac{\rho_o}{\rho_w} V_o$ tells us that the volume of fluid displaced is less when the object is in saltwater than when the object is in freshwater.

For option a to be true, the densities of freshwater and saltwater would have to be the same. For option b to be true, the density of saltwater would have to less than that of freshwater. Option d would require the density of the object to equal the density of the water. Option e does not apply to a situation in which an object floats.

28. Answer: d

The equation of continuity states that $A_1v_1=A_2v_2$. If $r_2>r_1$, $A_2>A_1$. This means that $v_1>v_2$. The speed of the fluid will decrease.

Option a, b, c, and e do not agree with the equation of continuity.

29. Answer: c

The equation of continuity states that $A_1v_1=A_2v_2$. If $v_1/v_2=4$, $A_2/A_1 = 4$. This means that $\pi r_2^2/\pi r_1^2 = 4$, or $r_2/r_1 = 2$. Therefore, $r_2 = 2r_1$.

Option a, b, d, and e do not agree with the equation of continuity.

30. Answer: a

The change in volume due to thermal expansion is approximately $3\alpha V\Delta T$.

$\Delta V = 50.05$ cm^3 − 50 cm^3 = 0.05 cm^3.

0.05 cm$^3 \approx 3\alpha V\Delta T$.

$\Delta T \approx 0.05$ cm$^3/3\alpha V = \frac{0.05}{3(1\times 10^{-5})(50)} = 33\ K$.

The initial temperature is 274 K. The volume will be 50.05 cm^3 at a temperature of approximately 307 K.

Options b-d are not the correct order of magnitude. Option e would not result in expansion because it is a lower temperature than the initial temperature.

31. Answer: e

Some real gases will exhibit this behavior. An ideal gas, however, has no interactions between molecules and will not become solid at low temperatures.

Options a-d are all true for ideal gases.

32. Answer: b

The First Law of Thermodynamics states that $\Delta U = Q - W$. If the system gains heat, Q is positive. If work is done by the environment on the system, W is negative. If Q is positive and W negative, ΔU must increase.

For option a, work done on the environment means that W is positive; the system gains heat, meaning that Q is positive. From the First Law of Thermodynamics, we see that ΔU= positive quantity − positive quantity. However, we don't know the values of Q or W, so we cannot determine the sign of ΔU. For option c, positive W and negative Q result in a negative value for ΔU. For option d is not correct, negative W, negative Q gives ΔU = negative value plus a positive value. However, we don't know the values of Q or W, so we cannot determine the sign of ΔU. For option e, negative W, unknown Q. We know Q changes but we cannot determine its value or sign. So, we cannot determine what happens to ΔU.

33. Answer: c

The definition of an adiabatic process is one for which no heat flows into or out of the system.

34. Answer: a

The work done is equal to the area under the curve. In this case, P is constant. The area is (50 kPa)(0.5 m^3)= 25 J.

35. Answer: b

The maximum efficiency is given by:

$$1 - \frac{T_c}{T_h} = 1 - \frac{200}{500} = 1 - 2/5 = 3/5 = 0.6.$$

Options a and c-e are not the correct order of magnitude.

36. Answer: c

The force between particles is f = kq$_1$q$_2$/r^2.
f = (9×10^9)(1×10^{-6})(2×10^{-6})/0.5^2 = 0.072 N

Option a, b, d, and e are not the correct order of magnitude.

37. Answer: a

We only need to calculate the magnitude of the charge, so we only need to use the magnitudes of E and F.

E = F/q.
q = F/E = 0.1/(1 × 10^4) = 10 µC

Option b, c, d, and e are not the correct order of magnitude.

38. Answer: e

An electric field of magnitude $k\frac{q}{r^2}$ is produced by a point charge. Since the point charge is negative the field id directed radially inward. Option e is the only option in agreement with the magnitude of the field and the direction.

39. Answer: d

The work done by the electric field is $W = -QEd$. The negative sign arises because the charge Q moves in the opposite direction to the electric field.

$$W = (-1 \times 10^{-6} \text{ C})(1 \times 10^4 \text{ N/C})(0.1 \text{ m}) = -0.001 \text{ J}$$

Options a and b are the wrong order of magnitude and do not take into account the sign. Option c has the correct magnitude but the wrong sign. Option e is the wrong order of magnitude.

40. Answer: a

$\Delta V = Ed$. Note that here the change in potential will be positive because the charge Q moves in the direction opposite to the electric field and so, the electric potential increases. $\Delta V = (1 \times 10^4 \text{ N/C})(0.1 \text{ m}) = 1 \times 10^3 \text{ V}$.

Options b and d have a negative value for ΔV. As described above this is not the case. Options c and e are not the correct order of magnitude.

41. Answer: b

Options a, c, d, and e all describe an ideal conductor. Option b is the only option not true of ideal conductors.

42. Answer: a

$C = \frac{\varepsilon_0 A}{d}$.

Rearranging, for A, we get:

$$A = \frac{Cd}{\varepsilon_0} = \frac{(1 \times 10^{-11})(1 \times 10^{-4})}{9 \times 10^{-12}} = 1 \times 10^{-2} \text{ m}^2.$$

Options b-e are not the correct order of magnitude.

43. Answer: b

From Ohm's Law, we know that V=IR. We can rewrite this as I=V/R. For a series circuit, the equivalent resistance is the sum of the individual resistances.

$$R_{eq} = R_1 + R_2 + R_3 + R_4 = 600 \Omega$$

The current that flows through a series circuit is the same through each resistor.

I = 20 V/600Ω = 1/30 A

Options a and c-e do not obey Ohm's Law.

44. Answer: e

The current through the circuit = 1/30 A. We can find the voltage drop across R_3 by using Ohm's Law V=IR.

$$V = \frac{1}{30} A \times 150 \, \Omega = 5 \, V$$

Option a is not the correct order of magnitude. Option b is the voltage supplied to the circuit. Options c and d do not follow Ohm's Law.

45. Answer: c

R_5 has a value of 500Ω. This is the same as $R_2 + R_3 + R_4$ in the original circuit. Therefore, both circuits have the same equivalent resistance and the current through both circuits is the same.

Options a, b, d, and e do not agree with Ohm's Law.

46. Answer: d

A voltmeter needs to be placed in parallel with a resistor to measure the voltage across the resistor. It also needs to draw as little current as possible. Therefore, an ideal voltmeter has infinite resistance.

Options a, b, and e are incorrect. An ideal ammeter needs to be placed in series with the resistor and would have zero resistance. Option c places the voltmeter in series, which as explained above, is not the correct placement for a voltmeter.

47. Answer: a

The capacitors are in series with each other. To obtain the equivalent capacitance, we add the two capacitances. $C_{eq} = 5 \times 10^{-10}$ F.

Option b subtracts the two capacitances. Option c is not correct; it treats the capacitors as resistors. Options d and e are the wrong order of magnitude.

48. Answer: d

To experience a force, the charge Q needs to move in a direction perpendicular to the direction of the magnetic field. A particle also needs to have a charge to experience a force due to the magnetic field. Option d is the only option satisfying these conditions.

In option a, the charge is stationary. No movement means that a force does not act on the charge. Options b and c involve movement in the y direction. Option e involves a neutral particle. As mentioned above, a neutral particle will not experience a force here.

49. Answer: d

The current and magnetic field are at right angles to each other, so the force on the wire is given by *ILB*. F=(5)(2)(1)= 10 N. To find the direction of the force, we use the right hand rule. Fingers point in the direction of the current, then curl towards the direction of the magnetic field. The thumb will then point in the direction of the force. In this case, the RHR tells us that the force is directed into the page.

Option a, b, and e specify a direction for the force that is not perpendicular to both the magnetic field and the current. Option c specifies the opposite direction to the actual force.

50. Answer: c

The fields from both wires will be in the same direction between the wires when the current travels in opposite directions. The strength will be higher between the wires when the two fields combine. Both options c and d have the current traveling in opposite directions. Option c, however, has larger currents flowing through the wires and so will produce larger magnetic fields than those in option d.

Options a, b, and e have the currents traveling in the same direction through the wires, and so the magnetic fields will be in opposite directions between the wires.

51. Answer: a

The flux is given by $BA\cos\theta$. (2)(0.1)cos60° = 0.1 Wb.

Options b-e have the wrong order of magnitude.

52. Answer: d

For a current to flow, the magnetic flux must change with time. One way to accomplish this is for the magnitude of the magnetic field to change with time.

Options a and b will not result in a current because the magnetic flux does not change. Option c and e have a magnetic flux of 0. Again, it does not change with time and therefore, current is not induced.

53. Answer: d

The magnetic field is directed upwards but the flux through the coil is decreasing (the magnetic field must be changing in time). A current will be induced in the coil. The current must flow in a direction so that the change is countered. Since the flux is decreasing through the coil, the current must flow counterclockwise. This will produce a magnetic field upwards, therefore, opposing the change.

Option b provides the wrong direction for the current. Options a, d, and e do not agree with Lenz's law.

54. Answer: c

A longitudinal wave is a wave in which the particles in the medium move in the same direction as the wave. A sound wave is an example of this.

Options a, b, and d are examples of transverse waves. Option e is more complicated; a water wave can produce a complicated motion.

55. Answer: d

The speed of a wave is given by $v = \frac{\lambda}{T}$ or λf. The speed is the same for both the original wave and the new wave, so the ratio $\frac{\lambda}{T}$ also remains the same. If λ is greater for the new wave, it follows that T must also be greater. This means that the frequency of the new wave is less than the frequency of the original wave.

56. Answer: a

For a string fixed at both ends vibrating in the first harmonic, the length of the string equals half the wavelength. In this case, the length of the string is 1 m. Therefore, the wavelength equals 2 m.

Options b, c, and e correspond to other harmonics. Option d is not possible for the wavelength. The largest possible wavelength corresponds to the first harmonic, which, in this case, has a wavelength of 2m.

57. Answer: b

The wavelength of the first harmonic is 2L for a wave on the string and 4L for a tube open at one end.

Options a and c-e do not accurately compare the first harmonics.

58. Answer: b

The amplitude of the wave is the maximum displacement from equilibrium. This can be found from the graph to be 0.5 m (the maximum distance away from the equilibrium position of 0 m).

Option a gives a value twice the amplitude. Options c and d are time values, not distance. Option e is the equilibrium position.

59. Answer: b

The waves are coherent so they can interfere constructively or destructively. To find the amplitude of the wave at point P that results from the interference of the two waves, we first need to find the path difference (the difference between the distances each wave traveled to point P). The difference in the path length is $\frac{5\lambda}{2} - \frac{\lambda}{2} = \frac{4\lambda}{2} = 2\lambda$. Since the path difference is equal to a multiple of the wavelength, the waves interfere constructively and the amplitude at point P will be the sum of the amplitude of the waves. In this case, 2A.

Options a and c-e do not correspond to constructive interference.

60. Answer: b

The law of reflection tells us that the angle of incidence is equal to the angle of reflection. Therefore, the angle of reflection is 40°.

Options a and c-e do not follow the law of reflection.

61. Answer: c

If the surface is rough, not smooth, the angle at which a ray will be reflected depends on where it strikes the surface. Rays will be deflected at different angles at different locations on the surface. The Law of Reflection does not hold.

62. Answer: c

The speed of light in a medium is given by $v = c/n$. If $\frac{n_1}{n_2} > 1$, $n_1 > n_2$. This means that $v_1 < v_2$. The speed of light through the second medium is greater than the speed through the first medium.

Option a and e are not correct – we have no information about the angle at which light enters the medium, so we cannot know if the light will be totally internally reflected. Although, the question does state that the light enters the second medium. This implies that total internal reflection does not occur. Option b is not correct - $\frac{n_1}{n_2} > 1$ tells us that $n_1 > n_2$ and $v_1 < v_2$. Option d is not correct – the index of refraction is different in both mediums, so the speed differs.

63. Answer: d

From Snell's Law, we know that $n_1 \sin\theta_1 = n_2 \sin\theta_2$. If $\theta_1 = 0$, then $n_1 \sin\theta_1 = 0$. This tells us that $n_2 \sin\theta_2$ must also equal 0. For this to occur, $\sin\theta_2 = 0$ which gives $\theta_2 = 0$.

Options a-c have an angle of incidence other than 0. From Snell's Law, we see that the light will be refracted as it enters the second medium. Option e is not true as discussed above.

64. Answer: d

A plane mirror is a mirror with a flat surface. Ray tracing can be used to show that the image is upright and virtual. We can think of what we see in our typical bathroom mirrors.

Options a and b describe the image as inverted, which we know from experience is not true. Option c describes the image as real, which again, we know is not true. Option e gives two true points, but the image formed is not bigger than the object, as can be seen from ray tracing.

65. Answer: c

The energy of a photon is given by E = hf. If the photon is a 'particle' of light from the visible spectrum, it's frequency is on the order of 5×10^{14} Hz. So, the energy ≈ $(6 \times 10^{-34})(5 \times 10^{14}) \approx 30 \times 10^{-20} = 3 \times 10^{-19}$ J. Converting this to eV (1 eV ≈ 1.5×10^{-19} J), we see that E≈2 eV. The energy is on the order of 1 eV.

Options a, b, d, and e are all the wrong order of magnitude.

66. Answer: d

This is essentially the definition of a photon.

Options a-c state that the rest mass of a photon is non-zero, which contradicts the statement that photons travel at the speed of light. Option e is correct for the rest mass but a photon does have a momentum when moving.

67. Answer: a

The de Broglie wavelength λ = h/p. Substituting in our values gives us a value of 7×10^{-7} m.

Options b-e are the wrong order of magnitude.

68. Answer: b

In the representation of a nucleus $^A_Z X$, A is the mass number (the number of nucleons) and Z is the atomic number (the number of protons). The representation of the sodium nucleus tells us that there are 22 nucleons and 11 protons in the nucleus.

Options a and c-e do not give correct values for A and Z according to this representation.

69. Answer: d

During alpha decay, the nucleus loses two protons and two neutrons. The atomic number gives us the number of protons. Therefore, the atomic number will decrease by 2.

Options a and b do not correspond to alpha decay. An alpha particle is released in alpha decay, not a photon or a positron. Options c and e are not in agreement with alpha decay.

70. Answer: a

The best estimate for the uncertainty due to a measurement instrument is ½ to 1/5 of the smallest scale division. Since the smallest division is 0.5 mm, the best estimate of the uncertainty from the available options is 0.2 mm. Options b and c are very close to the value of the smallest scale division and do not provide a good estimate of the uncertainty. Option d and e are not the correct order of magnitude.

Free Response Questions

1.

I. To find the range, the horizontal distance traveled by the object, we first find the time t the object is in the air. We can use the equation:

$$y = y_o + v_{oy}t - \frac{1}{2}gt^2.$$

$y_o = 0$, if we set the origin to be the initial location of the object.

$v_{oy} = v_0 \sin\theta.$

$y - y_o = 0$

(we are calculating the time it takes for the object to hit the ground).

Therefore, our equation becomes:

$$(v_0 \sin\theta)t = \frac{1}{2}gt^2.$$

Simplifying and substituting in the given values, $5(\sin 20°) = \frac{1}{2}(10)t.$

We can now solve for t. t = 0.34 s.

Next, we calculate the range using the equation:

$$x = x_o + v_{ox}t.$$

Again, x_o can be set to 0 and $v_{ox} = v_0 \cos 20°$.

$$x = (5\cos 20°)0.34 = 1.6 \text{ m}$$

II. The time it takes to reach the maximum height is half the time the object is in the air. We calculated the time in the last part to be 0.34 s. The time it takes the object to reach its maximum height is 0.17s.

III.
   a. $v_x = v_{ox}$
      $= 5\cos 20°$
      $= 4.7 \text{ m/s}$
   b. The x component of the velocity does not change because there is no acceleration in the x direction. The only acceleration occurs in the −y direction and is due to gravity.

2.

a.

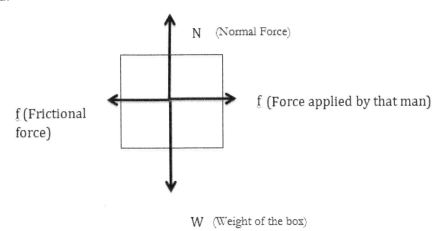

b. The object experiences a force of static friction that opposes the motion. The force is due to the roughness of the surface and the object. For the object to remain stationary the force of static friction must be equal but opposite to the applied force. It takes a force greater than the maximum value of static friction for the object to begin to move.

c. The maximum value of static friction is $\mu_s N$. If the applied force is larger than this, the object will begin to move.

From the free-body diagram, we see that the normal force is equal to the weight.

$$N = mg = (1\ kg)(10\ m/s^2) = 10\ N$$
$$\mu_s N = (0.5)(10\ N) = 5\ N$$

If the applied force is greater than 10 N, the object will start to move.

d. If the force applied by the man is removed, the object will have only two forces acting on it: the force due to static friction acting up the slope and the component of the object's weight along the direction of the slope.

If $mg\sin\theta > f_{smax}$, the object will slide down the slope. If $mg\sin\theta \leq f_{s,max}$ the object will stay still. (Continued on next page.)

$$f_{smax} = \mu_s N$$

Note that the normal force here is not simply the weight but rather the component of the weight perpendicular to the slope mgcosθ, the maximum value of:

$$f_s = \mu_s mg\cos\theta$$

Substituting in our values, we find that:

$$f_s = (0.5)(1)(10)\cos(45°) = 3.54 \text{ N}$$

Now, we compare the component of the weight acting down the slope to:

$$\text{mgsin}\theta = (1)(10)\sin(45°) = 7.07 \text{ N}$$

Since the component of the weight acting down the slope is greater than the maximum force due to static friction, the object will slide down the slope.

3.
- a. At equilibrium, the mass is at rest. The weight of the mass (mg) acts downward and the spring exerts an upwards force (kd) where d is the distance the spring has stretched. For the mass to be at rest, these forces must balance. Therefore, mg = kd. We can now solve for d:

$$d = mg/k$$
$$= (0.2)(10)/10$$
$$= 0.2 \text{ m}$$

- b. The period is given by:

$$T = 2\pi \sqrt{\frac{m}{k}}$$

We can substitute in our values:

$$T = 2\pi \sqrt{\frac{0.2}{10}} =$$

0.89 s

- c. The effect of increasing the force constant by a factor of 4 is shown below. The period is reduced by 2. This can also be seen from the equation for the period:

$$T = 2\pi \sqrt{\frac{m}{k}}.$$

If k is increased to 4k:

$$T = 2\pi \sqrt{\frac{m}{4k}} = \pi \sqrt{\frac{m}{k}}$$

Therefore, the period is half that of the original value.

4. We can use the known pressure at 0.1 m to find the density of the unknown liquid. The surface of the container is open to the atmosphere and so the pressure at the surface of the liquid is equal to atmospheric pressure (1.01×10^5 Pa). The pressure at a depth h below the surface of a liquid can be found by:

$$P = P_{at} + \rho gh$$

Since we know P and h, we can solve for ρ.

$$\rho = (P - P_{at})/gh$$
$$= (1.03 \times 10^5 - 1.01 \times 10^5)/(10 \times 0.1)$$
$$= 2000 \text{ kg/m}^3$$

Now that we know ρ, we can finish the problem:

$$P_2 = P_1 + \rho gh$$
$$P_2 = 1.03 \times 10^5 + (2000)(10)(1)$$
$$= 1.23 \times 10^5 \text{ Pa}$$

5. First, we define the +x direction to be to the right on our line and the −x direction to be to the left.

   To find the position at which q_3 should be placed such that the net force on it is 0, first note that q_1 and q_2 are both positive while q_3 is negative. This means that q_3 will experience a force in the −x direction due to q_1 and a force in the +x direction due to q_2.

   Next, we calculate the magnitudes of the forces acting on q_3. Define r_1 to be the distance of q_3 to q_1. The magnitude of the force on q_3 due to q_1 is kq_1q_3/r_1^2. The magnitude of the force on q_3 due to q_2 is $kq_2q_3/(0.5-r_1)^2$. We can now write:

   $$\frac{kq_2q_3}{(0.5-r_1)^2} - \frac{kq_1q_3}{r_1^2} = 0.$$

   Simplifying, this becomes:

   $$\frac{\sqrt{q_2}}{0.5-r_1} = \frac{\sqrt{q_1}}{r_1}$$

   $$r_1 = \frac{0.5}{\left(1+\frac{\sqrt{q_2}}{\sqrt{q_1}}\right)} = 0.22 \text{ m}.$$

   Therefore, the charge q_3 should be placed a distance 0.22 m to the right of q_1. From the origin, q_3 should be placed at 0.72 m to be in equilibrium.

6. a. $^{31}_{14}Si \rightarrow ^{31}_{15}P + e^-$

   b. The change in mass = 30.974 u – 30.975 u = - 0.001 u.
   We need to convert the units of the mass to MeV/c²:

   (0.001 u) = 0.001(931.5 MeV/c²) = 0.9315 MeV/c²

   E =(0.9315 MeV/c²)×c² = 0.9315 MeV

7.
   1. We can find the intensity a distance of 5 m away from the singer by considering the singer to be a point source. The intensity is given by:

      $$I = \frac{P}{4\pi r^2}$$

      $$= \frac{0.09\ W}{4\pi (5m)^2} = 2.9 \times 10^{-4}\ W/m^2$$

   2. To calculate the distance an observer would have to be to be unable to detect any sound from the singer, we use the fact that the lowest intensity we can hear is on the order of 10^{-12} w/m². To find an estimate for r, we can again use:

      $$I = \frac{P}{4\pi r^2}$$

      Thus,

      $$4\pi r^2 = \frac{P}{I}$$

      $$r = \sqrt{\frac{P}{4\pi I}} \sim \sqrt{\frac{0.09}{4\pi(1 \times 10^{-12})}} \sim 8.5 \times 10^4\ m$$

8. We are all familiar with the Doppler Effect, if not necessarily by that name. When we hear an ambulance approaching us, it sounds different that when it is moving away from us. This is just one example.

   Consider a source (such as the ambulance) moving towards an observer. The frequency with which the sound waves reach the observer is greater than if the source was stationary. A stationary source emits sound waves with a frequency f.

   If that source then starts to move, the source will start to "catch up" with the wave fronts ahead of it. The new wave fronts emitted by the source will be closer to each other than if the source remained stationary. The opposite is true behind the source. Here, the source is moving away from the wave fronts and so the new wave fronts are farther away from each other than they would be if the source was stationary.

   It can also be shown that if the source is stationary and the observer is moving, the Doppler Effect occurs. An observer moving towards a stationary source will "run into" the wave fronts. The relative velocity of the waves between the observer and the source is greater than if both the source and observer were stationary. The opposite is true if the observer moves away from the source. The observer, then, is "running away" from the wave fronts.

Quiz 2

Multiple Choice

1. A particle has a velocity u towards the east and at t=0, is acted upon by an acceleration directed towards the west. If D_a and D_b denote the magnitude of the displacement of the particle in the first 10 seconds and the next 10 seconds respectively, then:
   a) $D_a < D_b$
   b) $D_a > D_b$
   c) $D_a = D_b$
   d) $D_a <= D_b$
   e) Insufficient data

   Answer: (E)

   Explanation: The magnitude of acceleration is the determining factor in the given scenario, for depending upon it the body may either come to rest in the first 10 seconds and then reverse its direction of motion, coming back to the original position in the next 10 seconds or it might travel in a different manner dictated by the magnitude of acceleration, such as travelling with a constantly decreasing velocity throughout the motion without any change in direction. So option B, C and D all are possible cases but nothing can be said for sure. Option A on the other hand is not possible because the acceleration is opposite to velocity, thus the displacement in the first 10 seconds will be more than that in next 10 seconds or equal but never less.

2. **A ball is released inside an elevator going up with an acceleration, *a*. The acceleration of the ball when seen from the ground, after the release, will be:**
   a) a upwards
   b) (g-a) downwards
   c) g downwards
   d) (a-g) upwards
   e) g upwards

   Answer: (C)

   Explanation: The only force acting on the ball after the release is gravitational force. The acceleration of the ball due to this force is g downwards. Option B would have been correct when observations would have been made from the elevator itself. The other two options are completely invalid.

3. **A person is not able to see through fog because of:**
   a) The refractive index of fog is very high
   b) Phenomenon of total internal reflection occurs in fog droplets
   c) Absorption of light takes place
   d) Scattering of light by water droplets in fog
   e) None of the above

   Answer: (D)

   Explanation: Scattering of light occurs because the size of particles in fog is greater than the wavelength of incident light upon it. As a result light is scattered all around and nothing is visible. The other three phenomena do not occur while fog is formed. The refractive index is not responsible for blurred view. TIR takes place during the formation of rainbows.

4. If the height of the image is half the height of the object, formed in a concave mirror of focal length 40cm, then the position of the object in front of the concave mirror is:
   a) -40cm
   b) -80cm
   c) -120cm
   d) -100cm
   e) -160cm

   Answer: (C)

   Explanation: Magnification = - position of image/position of object:

   $-½ = -v/u \Rightarrow v = u/2$

   Also:

   $v = uf/(u-f)$

   Substituting, values of v, u & f in above equation we get:

   $(u/2) = u(-40) / u - (-40)$

   $u = -120$

   (Where, u = position of object, v = position of image and f = focal length of the concave mirror)

5. **The angle between two plane mirrors so as to obtain 3 images of a symmetrically kept single object is:**

    a) 30°
    b) 60°
    c) 90°
    d) 120°
    e) 180°

    Answer: (C)

    Explanation: When an object is placed between two mirrors, kept at an angle Θ, then multiple images of object are formed due to successive reflections.

    For an object placed symmetrically, the number of images formed are:

    n = (360/ Θ) – 1

    Therefore:

    3 = (360/ Θ) – 1

    On solving, we get:

    Θ = 90°

6. **The sum of all electromagnetic forces between different particles of a system of charged particles is zero:**

    a) Irrespective of the signs of the charges
    b) Only when all the particles are positive
    c) Only when all the particles are negative
    d) Only when there is an equal number of positive and negative charges
    e) Insufficient data

    Answer: (A)

    Explanation: Each particle exerts forces on the other particles which are equal in magnitude and opposite in direction to the forces applied on it by the neighboring particles. This is in accordance with Newton's Third Law of Motion. As a result (and irrespective of the signs of the charges) the sum of all the forces is zero. Options A, B and C are wrong because they vouch only for particular cases.

7. A block of mass 10 kg is suspended through two light spring balances as shown:

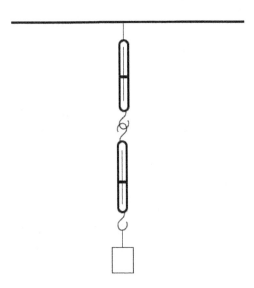

a) Both the scales read 5 kg
b) Both the scales read 10 kg
c) The upper scale reads 10 kg and the lower zero
d) Whatever the readings, their sum will be 10 kg
e) 7 kg and 3 kg

Answer: (B)

Explanation: The two spring balances at the equilibrium state act as a single string. Since a spring balance measures the tension acting through it and the tension in a string remains constant throughout, both the spring balances show equal readings. The reading here is equal to the weight of the block, which is equal to 10 kg. Hence, other options are not correct.

8. A man standing inside a stationary elevator drops a coin. He records that the coin reaches the elevator floor in a time t_1. When he repeats the same experiment with an elevator moving with a constant velocity, he records the time to be t_2. Then:

   a) $t_1 = t_2$
   b) $t_1 < t_2$
   c) $t_1 > t_2$
   d) It depends on whether the lift is going up or down
   e) Insufficient data

Answer: (A)

Explanation: Since the recordings are made by the man inside the elevator, all the observations are made from the frame of the elevator, which being in a non-accelerated motion, acts as an inertial frame. In this frame, during both the experiments the coin has the same initial velocity as that of the elevator frame, which can in turn be treated as a case of zero initial velocity of the coin. Also the coin undergoes acceleration only due to the gravity in both the cases, thus the time taken to reach the elevator floor is the same in both the cases. The options B, C and D do not hold, because no matter what, if the elevator is not accelerated, then the coin will have the same initial velocity as the elevator at the point of releasing, making it reach the floor in the same time interval for all cases.

9. The mass of a rider and his cycle combined is 90kg. The increase in Kinetic energy if the speed increases from 6 km/h to 12 km/h is:

   a) 300 J
   b) 350 J
   c) 375 J
   d) 390 J
   e) 400 J

   Answer: (C)

   Explanation: Let the mass of the rider be m_c and that of the bike be m_b.

   $M = m_c + m_b = 90$ kg

   $u = 6$ km/h $= 1.666$ m/sec

   $v = 12$ km/h $= 3.333$ m/sec

   Increase in K.E. =

   $\frac{1}{2} Mv^2 - \frac{1}{2} Mu^2$

   $= \frac{1}{2} \times 90 \times (3.333)^2 - \frac{1}{2} \times 90 \times (1.66)^2 = 494.5 - 124.6$

   $= 374.8$ (approx. 375 J)

10. Given four lenses having focal lengths +250cm, +175cm, +150cm, +20 cm and +15cm. The focal length of the eyepiece, to make an astronomical telescope, to produce largest magnification is:

    a) +15 cm
    b) +20 cm
    c) +150 cm
    d) +175 cm
    e) +250 cm

    Answer: (A)

    Explanation: Magnification in the case of an astronomical telescope is defined as the focal length of objective lens divided by the focal length of the eye piece. Magnification is inversely proportional to the focal length of the eye piece; hence, it should be small for greater magnification. The smallest of all options will be the best choice.

11. Two waves with the same intensity produce interference. If maximum intensity is 4 L, then the minimum intensity will be

    a) 0 L
    b) 1 L
    c) 2 L
    d) 3 L
    e) 4 L

    Answer: (A)

    Explanation: The intensity (I) is the same for both waves. Let amplitude be A.

    $$I = A^2$$
    $$I_{max} = (A+A)^2 + (2A)^2 = 4 A^2 = 4 L$$
    $$I_{min} = (A - A)^2 = 0$$

12. Given a tuning fork of frequency 256 Hz, it generates 4 beats per second with another tuning fork. The second tuning fork is now loaded with a little wax and the new beat frequency is found to be 6 beats per second. Find the original frequency of the tuning fork.

    a) 252 Hz
    b) 256 Hz
    c) 258 Hz
    d) 260 Hz
    e) 266 Hz

    Answer: (A)

    Explanation: The frequency of the unknown tuning fork can be either 256 - 4 = 252 Hz or 256 + 4 = 260 Hz. Since we apply wax to it, its mass/unit length increases. This implies that its frequency decreases. The current beats per second are now 6 so the original frequency must be 252 Hz, because 260 Hz is not possible (decreasing the frequency means the beats decrease, too.)

13. **A wave with a speed of 10 m/s is travelling on a string causing each particle to oscillate with a time period of 20ms. The wavelength of the wave is:**

    a) 10 cm
    b) 20 cm
    c) 30 cm
    d) 35 cm
    e) 40 cm

    Answer: (B)

    Explanation: Given that wave propagation speed is v = 10 m/sec, the time period T is equal to:

    20 ms = 20 x 10^{-2} sec

    The wavelength is:

    vT = 10 x 2 x 10^{-3} = 0.2m = 20cm

14. **If nodes are formed at a distance of 4.0 cm, in a standing wave pattern in a vibrating air column, the speed of sound in air is 328 m/s , what is the frequency of the source?**

    a) 3.9 Hz
    b) 4.0 Hz
    c) 4.1 Hz
    d) 4.3 Hz
    e) 4.2 Hz

    Answer: (C)

    Explanation: We that velocity is equal to wavelength multiplied by frequency.

    Wavelength = 2 x 4.0 = 8 cm

    Frequency = 328/(8 x 10^{-2}) = 4.1 Hz

15. A slit which is long, narrow and horizontal is placed 1 mm above a horizontal plane mirror. Interference is observed between the light coming directly from the slit and that after reflection, on a screen 1 m away from the slit. If the light used has a wavelength 700 nm, then fringe – width is:
    a) 0.30 mm
    b) 0.35 mm
    c) 0.38 mm
    d) 0.40 mm
    e) 0.45 mm

Answer: (B)

Explanation: Given that:

D (Distance of screen) = 1.0 m

W (Wavelength) = 700 x 10^{-9} m

d (Separation between the slits) = 2 x (1 x 10^{-3}) m = 2 x 10^{-3} m

Fringe Width = WD/d = 700 x 100^{-9}m x 1m / (2 x 10^{-3}) = 0.35 mm

16. Which of the following statements is not true about optical fibers?
    a) Works on the principle of total internal reflection
    b) have electro-magnetic interference from outside
    c) Have very low transmission loss
    d) Have high speeds of transmission
    e) Suitable cladding with homogenous core

Answer: (B)

Explanation: Electro-magnetic interference from outside is not possible due the protective coating that protects the optical fiber. Electro-magnetic interference creates noise in the signal that is being transferred from optical fiber which inevitably needs to be avoided. Thus option B is invalid. The other options are all true about an optical fiber.

17. Two statements are made:

    (I) The linear momentum of a particle is independent of the frame of reference.

    (II) The kinetic energy of a particle is independent of the frame of reference.

    About these statements:

    a) Both the statements are true
    b) Both the statements are false
    c) (I) is true, (II) is false
    d) (I) is false, (II) is true
    e) Can't be determined

    Answer: (B)

    Explanation: The linear momentum of a body is defined as the product of mass and velocity, whereas the kinetic energy is given by half of the product of mass and the square of speed of the body. Both the quantities (velocity and speed) depend on the frame of reference because a particle in motion in one frame may be at rest with respect to another frame (such as the frame attached to the particle itself). Thus both the statements are false since they are based on quantities that are dependent on frame of reference. Making options A, C, D and E invalid

18. A bullet travelling horizontally hits a block placed on a smooth horizontal surface and gets embedded into it. The quantity that <u>does not</u> change is:

    a) Linear momentum of the block
    b) Kinetic energy of the block
    c) Temperature of the block
    d) Gravitational potential energy of the block
    e) Speed of the block

    Answer: (D)

    Explanation: Since a high velocity bullet gets embedded into the block, it changes its linear momentum, kinetic energy (by increasing the speed) and temperature (due to local deformation during embedding). There is no change in the vertical height of the block, thus the gravitational potential energy remains constant. Momentum changes because of change in speed and mass (bullet gets embedded).

19. A lady is sitting on a rotating stool with her arms stretched out. If she folds her arms, then the magnitude of her angular momentum and angular velocity:

a) Both remain constant
b) Both change
c) Angular momentum remains constant, angular velocity changes
d) Angular momentum changes, angular velocity remains constant
e) Need actual to data to determine

Answer: (C)

Explanation: Since there is no external force involved, the momentum is conserved. The folding of her arms results in a decrease in the moment of inertia. In order to compensate the decrease in moment of inertia, so as to keep the overall angular momentum constant, the angular velocity of the object increases. Thus option A, B, D and E are incorrect.

20. An astronaut sitting in a chair in the International Space Station feels weightless because:

a) The earth does not attract the objects in the space station
b) The normal force by the chair on the astronaut balances the earth's attraction
c) The normal force is zero
d) The astronaut is not accelerated
e) No gravity in space station

Answer: (C)

Explanation: The earth does attract the objects in the space station, but the force is quite feeble. The astronaut is in an accelerated state since he, along with the entire space station, is orbiting the earth. But due to the absence of any normal force applied on him by the chair, he feels weightless. The only correct option is C.

21. A block is sliding on rough incline plane. The direction of friction acting on the surface of the incline is:

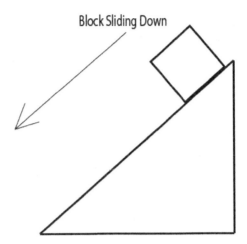

a) Perpendicular to the incline
b) Along the incline in the downwards direction
c) Along the incline in the upwards direction
d) Along the incline in both directions
e) Insufficient data

Answer: (B)

Explanation: The friction between any two surfaces always acts in way that opposes the relative motion of the surface. Here, the block is moving down with respect to the incline; as a result it experiences friction in the upward direction. The block, in turn, exerts a force of friction on the incline directed in the downwards direction along the incline. Thus only option B holds true.

22. A police car is following a culprit on a stolen superbike. The bike turns a corner at a speed of 72 km/h. The police follow at 90 km/h. Assuming both the vehicles are moving at constant speeds, if the police turn the corner 10 seconds after the superbike, what is the distance from the corner that the police will catch up with the culprit?:

a) 1000m
b) 2000m
c) 1500m
d) 1100m
e) 1200m

Answer: (A)

Explanation:

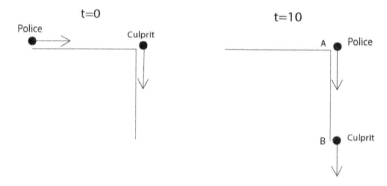

$V_P = 90$ km/h $= 25$ m/s.

$V_C = 72$ km/h $= 20$ m/s.

In 10 seconds, the culprit reaches point B from A. Distance covered is:

S = vt = 20 × 10 = 200 m.

In 10 seconds, the police jeep is 200 m behind the culprit (relative velocity considered) so the time is equal to:

s/v = 200 / 5 = 40 s

In 40 seconds, the police will move from A to a distance S, where:

S = vt = 25 × 40 = 1000 m

Thus, the police will catch up with the bike at 1 km from the turn.

23. An elevator is descending with an acceleration of 2 m/s². The mass of block A, which is placed on top of B, is 0.5 kg. The force exerted by A on B is:

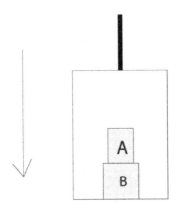

a) 8 N
b) 6 N
c) 4 N
d) 10 N
e) 30 N

Answer: (C)

Explanation:

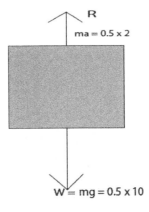

From the free body diagram:

$= R + 0.5 \times 2 - w = 0$

$= R = w - 0.5 \times 2$

$= 0.5 (10 - 2) = 4N.$

Therefore, the force exerted by block A on block B is 4N.

24. **Two cars with different masses m₁ and m₂ move in circles of different radii r₁ and r₂ respectively. If they complete the circle in equal time, the ratio of their angular speeds ω₁/ω₂ around the center of their circular tracks is:**
    a) m_1/m_2
    b) m_2/m_1
    c) r_1/r_2
    d) 1
    e) 1/2

    Answer: (D)

    Explanation: Angular speed is nothing but the angle covered in unit time. Since both of them complete a circle in the same time, it means they cover 360° in equal time, have similar angular velocities. Thus the ratio is 1, which corresponds to option D only.

25. **The ratio of the linear momentum of two particles, if their kinetic energy is the same, is:**
    a) 1 : 4
    b) 1 : 3
    c) 1 : 2
    d) 1 : 5
    e) 2 : 1

    Answer: (C)

    Explanation: Let the mass of the particles be $m_1=1$ kg and $m_2=4$ kg respectively.

    According to the situation:

    $½ m_1v_1^2 = ½ m_2v_2^2$

    $\Rightarrow m_1/m_2 = v_2^2/v_1^2 \Rightarrow v_2/v_1 = \sqrt{(m_1/m_2)}$

    $\Rightarrow m_1v_1/m_2v_2 = (m_1/m_2) \times \sqrt{(m_2/m_1)}$

    Putting in the values of m_1 and m_2, we obtain the ratio 1:2.

26. A boy throws a ball up in the air vertically and then catches it again. During the entire flight, the total mechanical energy of the ball (mechanical energy being defined as the sum of kinetic and potential energy), will be:

a) Remains constant
b) Changes continuously
c) Insufficient data
d) First increases then decreases
e) First decreases then increases

Answer: (A)

Explanation: In the absence of external forces and when all internal forces (like gravity) are conservative in nature, the total mechanical energy of the system remains conserved. In ascent flight, the kinetic energy decreases and changes into gravitational potential energy, whereas in the decent flight, the potential energy decreases and changes into kinetic energy, but at any instant, their sum, mechanical energy, remains constant. Hence only option A is valid.

27. In a gravity-free hall, a tray of mass M carrying an ice cube of mass m is placed, with edge L at rest in the middle. When the ice melts, the distance by which the center of mass of the "tray plus ice" system descends is:

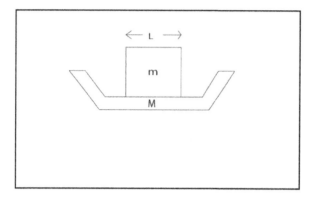

a) L/2
b) L/3
c) L/4
d) Zero
e) L

Answer: (D)

Explanation: Since the hall is gravity free, the melting cube of ice would tend to acquire a spherical shape, but would flow down entirely (no gravity). Also since there is no external force, the COM of the system remains in place. Thus, there is no change in position.

28. A dog chases a cat with a speed of 30 m/s for the first 10 seconds. Then it accelerates at a rate of 3 m/s² for the next 5 seconds before giving up entirely and coming to a stop. The total displacement of the dog in the entire 15 seconds-chase is:
    a) 300 m
    b) 187.5 m
    c) 487.5 m
    d) 0m
    e) 100m

    Answer: (C)

    Explanation: To get the displacement of the dog, we need to break the motion into two segments, 10 seconds non-accelerated chase and 5 seconds accelerated chase.

    Non-accelerated:

    10s x 30 m/s

    300 m

    Accelerated:

    u x t = 5s x 30m/s

    150 m

    $1/2(a \times t^2) = 0.5 \times 3 \times 25$

    37.5 m

    $s = ut + 1/2(at^2) = 187.5$

    Total area= displacement=

    300m + 187.5m

    487.5m

29. Part *a* of the following diagram shows a block suspended from a pulley. Now, the tension in the rope is 80 N. Part *b* shows the block being pulled up at a constant velocity, The tension in the rope in part *b* is:

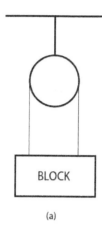
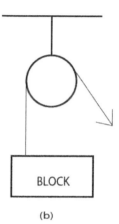

(a) (b)

a) 80 N
b) 160 N
c) 0 N
d) 100 N
e) 40 N

Answer: (B)

Explanation: In part (a) the equilibrium condition can be written as:

$2 \times T_a = mg$

In part (b) we have the motion with constant velocity. This implies that the acceleration is 0 and the net force is 0, which can be written as:

$T_b = mg$

From these two equations we can find:

$T_b = 2 \times T_a = 2 \times 80 N$

$= 160 N$

30. A small block with a mass of 0.1 kg lies on a fixed inclined plane PQ (P is the tip and Q is the bottom end of incline) which makes an angle θ with the horizontal. Now, a horizontal force of 1 N acts on the block through its center of mass, as illustrated in in the figure. The block remains stationary if: (g = 10 m/s²):

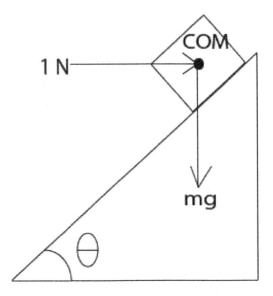

a) Only when θ = 45°
b) θ > 45° and a frictional force acts on the block towards Q.
c) θ >= 45° and a frictional force acts on the block towards P.
d) θ < 45° and a frictional force acts on the block towards P.
e) Insufficient data

Answer: (C)
Explanation:

At θ = 45°, mg sin θ = 1 cos θ (weight balances the external force)
At θ > 45°, mg sin θ > 1 cos θ (friction acts upward, towards P)
At θ < 45°, mg sin θ < 1 cos θ (friction acts downward, towards Q)

These are the three situations in which the block remains stationary, and this coincides with only option C. Hence the other options are incorrect.

31. The mass M = 9 kg and m = 3 kg. Given that the table is smooth, the acceleration of the block and the ball, once they are released from rest is:

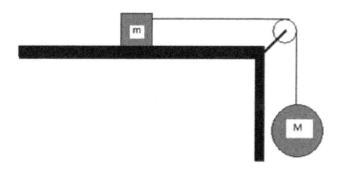

a) m/s²
b) m/s²
c) 6 m/s²
d) 7.4 m/s²
e) m/s²

Answer: (D)
Explanation: The tension of the cord is the same everywhere on the cord. The accelerations of both the masses are the same.

$$T = ma$$

$$Mg - T = Ma$$

$$Mg - ma = Ma$$

$$a = Mg \div (m + M)$$

$$a = (9 \times 9.8) \div (9 + 3)$$

$$a = 7.4 \, m/s^2$$

32. A boy pulls a mass of 1 kg suspended from a spring over a distance of 0.1 m from the equilibrium state at B to C. The spring constant being 300 N/m, the elastic potential energy and kinetic energy at C is:

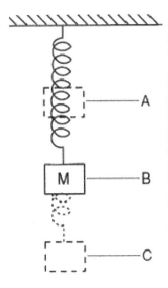

a) J and 1.50 J
b) J and 1.00 J
c) 0 J and 3.00 J
d) J and 0 J
e) 0 J and 0 J

Answer: (D)
Explanation: Spring potential energy is given by the equation:

$$U_S = 1/2\ kx^2.$$

We plug in the given values and get:

$$U_S = 1/2\ (300)\ (.1)^2 = 1.5\ J$$

Since the block isn't moving, the kinetic energy is zero.

33. A pendulum clock consists of a rod connected to a small pendulum. If it is designed to show correct time at 20°C, how fast or slow will it go in 24 hours at 40°C: (Coefficient of linear expansion of rod is 1.2×10^{-6} /°C)
    a) 7.1 s
    b) 8.2 s
    c) 10.2 s
    d) 15 s
    e) 12 s

Answer: (C)

Explanation: The time period at temperature t is:

=> $T = 2(i_t/g)^{1/2}$

=> $T = 2(i_0(1+at)/g)^{1/2}$

=> $T = T_0(1 + 1/2 \, at)$

=> $T_{20} = T_0(1 + 1/2(20°C))$

=> $T_{40} = T_0(1 + 1/2(40°C))$

=> $T_{20}/T_{40} = 1 + (10°C)$

=> $(T_{40} - T_{20})/T_{20} = 10°C \times a = 1.2 \times 10^{-4}$

=> Difference in t = 24 hrs x 1.2×10^{-4}

=> 10.2 s

34. An iron rail track is laid in winter, when the average temperature is 18°C. The track consists of 12m sections placed one after the other. How much gap should be there between two sections so that there is no compression during summer when the maximum temperature goes to 48°C? (Coefficient of linear expansion of iron= 11x10^{-6} /°C)
    a) 0.4 cm
    b) 1 cm
    c) 0.8 cm
    d) cm
    e) 2 cm

Answer: (A)

Explanation:

L_0(length) = 12 cm, a(coefficient of iron) = 11 × 10–5 /°C

t_w(current temp) = 18°C, t_s(final temp) = 48°C

$L_w = L_0(1 + at_w)$ = 12 (1 + 11 × 10–5 × 18) = 12.002376 m

$L_s = L_0 (1 + at_s)$ = 12 (1 + 11 × 10–5 × 48) = 12.006336 m

Change in length =

12.006336 – 12.002376 = 0.00396 m = 0.4cm

35. The characteristics of a gas resembles more closely to that of an ideal gas at:
   a) Low pressure and low temperature
   b) Low pressure and high temperature
   c) High pressure and low temperature
   d) High pressure and high temperature
   e) None of the above.

Answer: (B)

Explanation: At low pressure and high temperature the gases tend to behave more as an ideal gas because intermolecular attraction is brought down considerably. High pressure and low temperature both bring the molecules closer resulting in increased intermolecular interactions resulting in deviation from ideal gas behavior, hence only option B is correct.

36. The Root Mean Square (RMS) speed of Oxygen at 300k is:
   a) 483 m/s
   b) 500 m/s
   c) 300 m/s
   d) 720 m/s
   e) 450 m/s

Answer: (A)

Explanation:

$$v_{rms} = (3RT/M_0)^{1/2}$$

$\Rightarrow \quad ((3 \times 8.3 \times 300)/0.032)^{1/2} = 483 \text{ m/s}$

37. Two grams of hydrogen gas are sealed in a vessel of volume 0.02 m³ and maintained at a temperature of 300 K. The pressure exerted by the hydrogen molecules on the walls of the vessel is:
   a) 1.23×10^5 Pa
   b) 2.23×10^5 Pa
   c) 3.23×10^5 Pa
   d) 6.23×10^5 Pa
   e) 7.23×10^5 Pa

Answer: (a)

Explanation:

$m = 2$ g, $V = 0.02$ m3 $= 0.02 \times 10 6$ cc $= 0.02 \times 10 3$ L

$T = 300$ K, $M = 2$ g,

$PV = nRT$

$PV = mRT/M$

$P \times 20 = (0.082 \times 300)/20$

$P = 1.23$ atm $= 1.23 \times 10 5$ pa $\approx 1.23 \times 10^5$ pa

38. The increase in the internal energy of 10g of water when it is heated from 0°C to 100°C and converted into steam at 100 kPa, is:
(density of steam being 0.6 kg/m³, Specific heat capacity of water = 4200 J°C/kg, Latent heat of vaporization of water being 2.5x 10⁶ J/kg)

a) .75 x 10⁴ J
b) .75 x 10⁴ J
c) .75 x 10⁴ J
d) .25 x 10⁴ J
e) .75 x 10⁴ J

Answer: (B)

Explanation: Mass = 10g = 0.01kg.

$P = 105 \text{ Pa}$

$dQ = Q_{H2o}\ 0° - 100° + Q_{H2o} - \text{steam}$

$= 0.01 \times 4200 \times 100 + 0.01 \times 2.5 \times 106 = 4200 + 25000 = 29200$

$dW = P \times dV$

change=(0.001/6.0) - (0.001/1000)

$= 0.01699$

$dW = PdV = 0.01699 \times 105\ 1699J$

$dQ = dW + dU$ or $dU = dQ - dW = 29200 - 1699 = 27501 =$

2.75×10^4 J

39. A gas is initially at a pressure of 100 kPa and its volume is 2.0 m³. Its pressure is kept constant and volume is increased by 0.5 m³. Now, its pressure is changed from 100 kPa to 200 kPa while keeping volume constant. The gas is brought back to its initial state with the pressure varying linearly with volume. The amount of heat supplied or extracted is:
a) 5 kJ
b) 0 kJ
c) 5 kJ
d) 0 kJ
e) 5 kJ

Answer: (C)

Explanation:

$$P_1 = 100 \text{ KPa}$$

$$V_1 = 2 \text{ m3}$$

$$dV_1 = 0.5 \text{ m3}$$

$$dP_1 = 100 \text{ KPa}$$

We find that area under expansion is greater than area under compression, so heat is extracted from the system. Amount of heat = Area under the entire curve.

$$= \frac{1}{2} (5 \times 10^4) \text{ J}$$

$$= 25 \text{ KJ}$$

40. In a cylindrical tube AB, water enters through end A with speed v_1 and leaves through end B with velocity v_2. The tube is always filled with water. In case 1, the tube is horizontal, in case 2, the tube is vertical with end A at lower position, in case 3, the tube is vertical with end A upward. The two velocities v_1 and v_2 will be equal when:

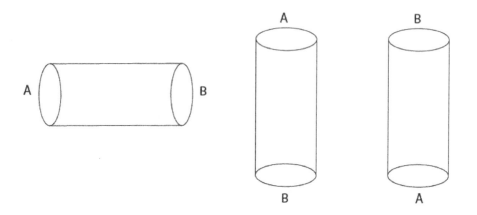

a) Case 1
b) Case 2
c) Case 3
d) All the above cases.
e) None of the above cases.

Answer: (D)

Explanation: On the application of Bernoulli's rule to all the three cases, we observe that in case 1, both the ends have equal pressure because both the ends are on the horizontal plane, i.e, there is no difference in height between the two ends, hence both ends have the same velocity. In case 2 and case 3, there is a change of pressure between the two levels, but the acceleration due to pressure is nullified by the effect of the potential difference between the two points. Hence, in this case also there will also be no change in velocity.

41. During nuclear fission:
a) A heavy nucleus splits up into lighter fragments by itself
b) A light nucleus breaks down due to bombardment by neutrons
c) A heavy nucleus breaks down due to bombardment by neutrons
d) Two or more light nuclei fuse to give a heavier nucleus and other products
e) New elements other than the reactant element are formed as products

Answer: (C)

Explanation: Nuclear fission is the process of breaking down a large nucleus into smaller fragments. The heavy nucleus has greater rest mass energy than those small fragments; this makes the process feasible (acquiring a state of lower energy). Thus, the heavy nucleus gets split into small fragments. Option A is incorrect because the process is not spontaneous. Option B involves a light nucleus, which is not true. Option D describes the process of nuclear fusion, which is the opposite of fission. Option E is incorrect, because the broken fragments have different atomic numbers than the original element.

42. What is the maximum energy a beta particle can have as a result of the following decay?

^{176}Lu -> ^{176}Hf + (other products)
Atomic mass of ^{176}Lu is 175.942694u and that of ^{176}Hf is 175.941420u.
a) 1.182 MeV
b) 2.122 MeV
c) 1.563 MeV
d) 1.231 MeV
e) 1.652 MeV

Answer: (A)

Explanation: The kinetic energy available for the beta particle is:

$$E = mc^2$$

$$E = [m(^{176}Lu) - m(^{176}Hf)]c^2$$

$$= (175.942694 \text{ u} - 175.941420)(931 \text{ MeV})$$

$$= 1.182 \text{ MeV}$$

This energy is shared by the beta particle and the antineutrino. The maximum kinetic energy of beta particle in the decay is, 1.182 MeV when the antineutrino particle does not get any share.

43. What is the maximum wavelength of a light that can ionize a hydrogen atom in its ground state? (The threshold energy of Hydrogen is 13.6 eV.) Round to the nearest whole number.
    a) 50 nm
    b) 70 nm
    c) 91 nm
    d) 86 nm
    e) 100 nm

Answer: (C)

Explanation: To ionize a hydrogen atom in its ground state, minimum of 13.6 eV is required, or in other words, the photon of light must have this much amount of energy to ionize it. Thus,

$$hC/\lambda = 13.6 \text{ eV}$$

$$\lambda = 1242 \text{ eV-nm}/(13.6 \text{ eV}) = 91.3 \text{ nm}$$

44. A parallel beam of monochromatic light (wavelength = 500nm) is incident normally on a perfectly absorbing surface. The power through any cross section of beam is 10 W. The number of photons absorbed per second by the surface is:
   a) 1.0×10^{19}
   b) 1.5×10^{19}
   c) 2.5×10^{19}
   d) 2.0×10^{19}
   e) 1.8×10^{19}

   Answer: (C)

   Explanation: The energy of each photon is:

   $$E = hC/\lambda$$

   $$= (4.14 \times 10^{-15})(3 \times 10^{08})/(500 \text{ nm})$$

   $$= 2.48 \text{ eV}$$

   In one second, 10 J of energy passes through the cross section of the beam. Thus number of photons crossing the cross section is:

   $$n = 10 \text{ J}/(2.48 \text{ eV}) = 2.52 \times 10^{19}$$

   Rounding off the data, the number of photons absorbed by the perfectly absorbing surface $= 2.5 \times 10^{19}$

45. The figure shows a wire of length L that can slide over a U-shaped rail of negligible resistance. Let the resistance of wire be R. The wire is pulled towards the right with a constant speed V. The current flowing in the wire is:

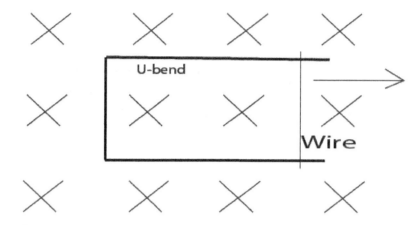

a) 2VBL/R
b) VBL/R
c) 4VBL/R
d) 3VBL/R
e) 8VBL/R

Answer: (B)

Explanation: The magnitude of the emf induced due to the moving wire is E=VBL. As the wire moves towards right, the force F= q(VxB) acts on the positive charge in upward direction of the plane. So the positive terminal of the equivalent battery appears upward and the resistance of the wire acts as the internal resistance of the battery.

The current flowing through the wire will be E/R, so the current = VBL/ R.

46. **A charged body moves in a uniform magnetic field with a constant velocity. At some instant of time the velocity vector makes an acute angle with the field vector. Then the path of motion of that body will be:**
    a) Circular
    b) Straight line motion
    c) Helix with uniform pitch
    d) Helix with non-uniform pitch
    e) Parabolic

    Answer: (C)

    Explanation: The velocity vector makes an acute angle with the field vector. One component of velocity is perpendicular to the magnetic field, while the other one is parallel to it. The force applied by the magnetic field is only on that component that is perpendicular to it (Cross vector product), So this force helps to change the motion to the circular path, while the other component of velocity, which is parallel to the magnetic field, experiences no force, and results in straight line motion with uniform speed. The superposition of the two motions (the circular and the straight line uniform motion) results in the formation of a helix with uniform pitch.

47. **A charged particle moving in gravity-free space undergoes deflection. Which of the following is false?**
    a) The electric and magnetic fields are both zero.
    b) The electric and magnetic fields are both non-zero.
    c) The electric field is non-zero, the magnetic field may be zero.
    d) The electric field is zero, the magnetic field is non-zero.
    e) Insufficient data

    Answer: (A)

    Explanation: Since the particle is moving in gravity-free space, no gravitational force is applied on it, but even then, the body undergoes deflection. So there must be some force that results in this deflection, which can be either magnetic or electric in nature. If both the magnetic and electric field is zero, then there would be no external force acting on the body, leading to no deflection. However, this is contradictory to the observation. Thus, option a is false.

48. Two resistances, R and 2R, are connected in series in an electric circuit. The ratio of the thermal energy developed in the two resistors, R and 2R, is:
   a) 1:2
   b) 2:1
   c) 1:4
   d) 4:1
   e) 1:5

Answer: (A)

Explanation: The thermal energy developed is equal to i^2R. Here, the current given is constant, so the thermal energy produced is directly proportional to the resistance. Hence, the ratio of thermal energy developed is 1:2.

49. A capacitor has the following features:
Distance between the plates = 10 cm
Area of Plate (A) = 2 m²
Charge on each plate (Q) = 8.85 x 10^{-10} C
What is the electric field between the plates?
(ε_0=8.854 x 10^{-12} C²N^{-1}m^{-2})
   a) 20 N/C
   b) 30 N/C
   c) 40 N/C
   d) 50 N/C
   e) 60 N/C

Answer: (D)

Explanation: As we know that the electric field outside the plates is zero, the electric field inside the plates is given by:

$$E = Q/\varepsilon_0 A = 50 \text{NC}^{-1}$$

50. The resistance of a copper wire of length 10 m, and cross-sectional area of 1.0 mm² is: (Resistivity of copper is 2.6 x 10⁻⁸ Ω-m)
    a) 1Ω
    b) 1.7Ω
    c) 0.5Ω
    d) 1.2Ω
    e) 1.3Ω

    Answer: (B)

    Explanation: Resistance (R)= Resistivity x(length/ Area)

    $= 1.7 \times 10^{-8}$ Ω-m x 10 m/(1.0 x 10⁻⁶ m²) = 0.17 Ω

51. A resister develops 400 J of thermal energy in 10s, when a current of 2 A is passed through it. What is the value resistance?
    a) 8 Ω
    b) 10 Ω
    c) 12 Ω
    d) 14 Ω
    e) 15 Ω

    Answer: (B)

    Explanation:

    $U = i^2Rt$

    400 J = (2 A)² x R x 10s

    R= 400/ 40

    = 10Ω

52. **Two metal plates with charge Q and -Q on opposite faces (like that of a capacitor), with some separation between them, are dipped into an oil tank. Now, if we pump out the oil, then, the electric field between the plates will:**
   a) Decrease
   b) Increase
   c) Become zero
   d) Increases then decreases
   e) Insufficient data

   Answer: (B)

   Explanation: The electric permeability of free space (or air) is greater than that of oil. As a result, when the oil gets removed from between, the strength of the electric field increases. This is proved only by option (B).

53. **A parallel plate capacitor has an area of 25 cm², and a plate separation of 2 mm. The capacitor is connected to a potential difference of 12 V. The charge on the capacitor is:**
   a) 4.32×10^{-10} C
   b) 3.32×10^{-10} C
   c) 2.32×10^{-10} C
   d) 1.32×10^{-10} C
   e) 5.32×10^{-10} C

   Answer: (D)

   Explanation:

   Plate area A = 25 cm² = 2.5×10^{-3} m

   Separation d = 2 mm = 2×10^{-3} m

   Potential v = 12 v

   Since we know that:

   $C = \varepsilon_0 A/d$ {$\varepsilon_0 = 8.85 \times 10^{-12}$}

   $C = (8.85 \times 10^{-12} \times 2.5 \times 10^{-3})/(2 \times 10^{-3}) = 1.06 \times 10^{-12}$ F

   $C = q/V = 11.06 \times 10^{-12} / 12 = 1.32 \times 10^{-10}$ C

   $Q = 1.32 \times 10^{-10}$ C

54. A capacitor stores 50×10^{-6} C of charge when connected across a battery. When the gap between them is filled with a di-electric, a charge of 100×10^{-6} C flows through the battery. The di-electric constant of the material inserted between the capacitors is: (Round off where ever required)

a) 1
b) 2
c) 3
d) 4
e) 5

Answer: (C)

Explanation: Initial charge stored = 50×10^{-6} C. Let the dielectric constant of the material induced be 'k'. The extra charge flown through the battery is 100×10^{-6} C. So, the net charge stored in the capacitor is:

150×10^{-6} C

Now, we can perform one of the following:

$C_1 = \varepsilon A/d$

$Q_1/V = \varepsilon A/d$ _(1)

So we get one of two answers:

$C_2 = k\varepsilon A/d$

$Q_2/V = k\varepsilon A/d$ _(2)

Dividing (1) and (2), we get:

$Q_1/Q_2 = 1/k$

$50/150 = 1/k$

$k = 3$

55. The equivalent resistance between points a and b of the network shown is:

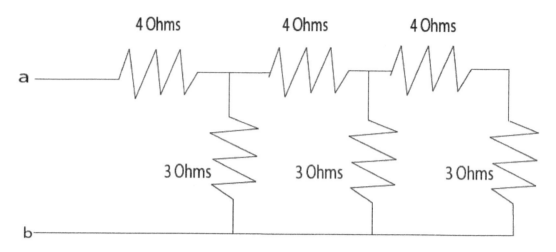

a) 4 Ohms
b) 6 Ohms
c) 7 Ohms
d) 9 Ohms
e) 8 Ohms

Answer: (B)

Explanation: The two rightmost resistors 3 Ω and 4 Ω are joined in series and may be replaced by a single resistor of 7 Ω. This 7 Ω is connected with adjacent 3 Ω resistor in parallel. The equivalent resistance of these two is (7 Ω x 3 Ω)/(7 Ω + 3 Ω) = 2.1 Ω. This is connected in series with adjacent 4 Ω resistor giving an equivalent resistance of 6.1 Ω which is connected in parallel with 3 Ω resistor. Their equivalent resistance is 2.01 Ω which is connected in series with the first 4 Ω resistor from left. Thus the equivalent resistance becomes 6 Ω (approximately).

56. The equivalent resistance between points a and b of the network shown is:

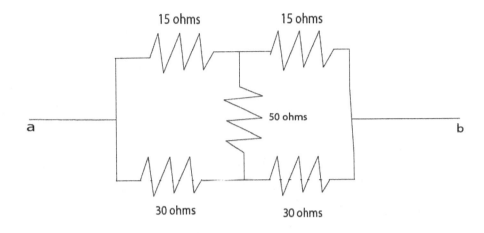

a) 10 Ω
b) 20 Ω
c) 30 Ω
d) 40 Ω
e) 50 Ω

Answer: (B)

Explanation: The circuit is equivalent to a wheat-stone bridge with galvanometer replaced by a 50 Ω resistor. As the bridge is balanced (R1/R2 = R3/R4), there will be no current through the 50 Ω resistor. The circuit is then equivalent to two resistors of 30 Ω and 60 Ω connected in parallel. The equivalent resistance is R=(30 Ω x 60 Ω)/(30 Ω + 60 Ω)=20 Ω.

57. The equivalent resistance of the following circuit is:

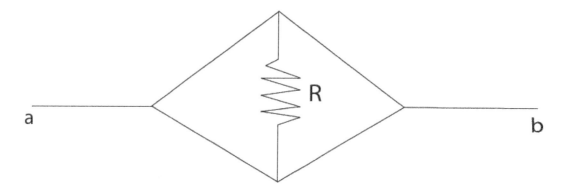

a) 1 Ω
b) 2 Ω
c) 3 Ω
d) 0 Ω
e) 4 Ω

Answer: (D)

Explanation: Current always flows through the path of least resistance. The central resistance has no current flowing through it because it offers resistance when there are already zero resistance paths available (the wire routes without any resistance). As a result, the circuit has an equivalent resistance of zero ohms.

58. If a projectile explodes into three pieces at the highest point of its trajectory, the center of mass of the three-particle system lands at:
a) The exact position of the landing of the intact projectile.
b) A bit further than the actual landing point.
c) A bit before the actual landing point.
d) At the midpoint of trajectory.
e) Insufficient data.

Answer: (A)

Explanation: The COM of the projectile travels along the original trajectory up to the highest point. At the highest point, the projectile explodes, since no external forces are involved and there are only internal forces at work, the COM undergoes no additional acceleration due to the explosion and carries on the original trajectory, thus landing at the original point determined by the range. Hence, only option A is correct.

59. When a cubic block is completely immersed in a vessel filled up to the brim with a liquid, the amount of liquid that overflows:
   a) Has equal weight to that of the block
   b) Has equal volume to that of the block
   c) Has the same relative density as that of the block
   d) Has the same mass as that of the block
   e) Cannot be determined

Answer: (B)

Explanation: The amount of liquid displaced is equal to the volume of the block that is immersed in the liquid. It has no relation with the mass or weight. Hence, only option B is correct.

60. Which of the following has the highest kinetic energy?
   a) A ball moving with a speed v
   b) A ball moving with speed v and rotating with an angular speed ω
   c) A ball rolling with a speed v on a rough surface
   d) A ball that is placed at rest at height h
   e) Insufficient data to decide

Answer: (B)

Explanation: A linearly moving and rotating ball has both translational kinetic energy and rotational kinetic energy. This is greater than simple translational kinetic energy. Hence, option B has maximum kinetic energy.

61. A block of mass 5 kg slides down an incline of inclination 30° and length 10 m. The work done by gravity is:
   a) 230 J
   b) 235 J
   c) 240 J
   d) 245 J
   e) 250 J

Answer: (D)

Explanation: Mass = 5 kg, angle of inclination = 30°, incline length = 10 m

Height = 5 m

Force = mass x gravity

Work done = mass x gravity x height = 5 x 9.8 x 5 = 245 J

62. An engine needs to lift 2000 kg of metal a distance of 12 m in 1 minute. The minimum horsepower of the engine needs to be: (1 hp = 746 W)

a) 5.1 hp
b) 5.2 hp
c) 5.3 hp
d) 5.4 hp
e) 5.7 hp

Answer: (C)

Explanation: Mass = 200 kg, distance = 12 m, time = 1 min = 60 sec

So:

work W = force x distance x cosθ = mass x gravity x distance x cos 0°

(θ = 0°, for minimum work)

= 2000 x 10 x 12 = 240000 J

Remember that:

power P = Work/time = 240000/60 = 4000 watt

Thus,

Horse Power = 4000/746 = 5.3 hp

63. From on top of a 40 m high cliff, a projectile is fired with an initial speed of 50 m/s at an unknown angle. Its speed when it hits the ground is:
    a) 57.4 m/s
    b) 65.5 m/s
    c) 52.3 m/s
    d) 22.5 m/s
    e) 33.5 m/s

    Answer: (A)

    Explanation:

    Height = 40m, initial speed u = 50 m/sec

Let the speed be 'v' when it strikes the ground. Applying the Law of Conservation of Energy:

Initial energy = final energy

Potential energy + initial kinetic energy = final kinetic energy

Mass x g (9.8) x height + ½ mass x u² = ½ mass x v²

⇨ 10 x 40 + (1/2) x 2500 = ½ v² => v² = 3300 => v = 57.4 m/sec

64. Which of the following is true?
   a) Momentum is conserved in all collisions but kinetic energy is conserved only in elastic collisions
   b) Momentum is conserved in all collisions but not kinetic energy
   c) Both momentum and kinetic energy are conserved in all collisions
   d) Neither momentum nor kinetic energy is conserved in elastic collisions
   e) Momentum does not depend on speed of the body

Answer: (A)

Explanation: No deformation occurs in elastic collisions, thus kinetic energy is conserved. Otherwise kinetic energy is not conserved in inelastic collisions, as a part of the energy is converted into heat and sound energy, and deformation in the form of potential energy. However, momentum is conserved in all types of collision. This corresponds to option A.

65. Newton's third law states that for every action there is an equal and opposite reaction. If team A pulls team B with equal magnitude of force in a game of tug of war, how does team A win?
   a) Tension is different in rope for the teams.
   b) The tension is same but the friction between the team and the ground is different.
   c) This game defies physics.
   d) None of the above.
   e) The rules are partial.

Answer: (B)

Explanation: Tension is always constant in a rope, but the friction that exists between the teams and the ground is different. Thus the difference in the magnitude of friction causes one team to win and another to lose. Other options are not valid reasons for the scenario.

66. A closed cylindrical container is half filled with oil, and all the air is pumped out from a small opening near the top of the container. What happens next?
   a) The force by the oil on the bottom of the vessel will decrease.
   b) The oil level will rise up in the vessel.
   c) The density of the liquid will decrease.
   d) The force will not change.
   e) The force by the oil on the bottom of the vessel will increase.

Answer: (A)

Explanation: The force at the bottom of the vessel is due to the combined pressure of the atmosphere and the oil's pressure due to its weight. As all the air is sucked out, there is no air in the vessel, resulting in zero atmospheric pressure. Thus, the force applied by air on oil is non-existent and the net force on the bottom of the container is only due to the oil's weight. Thus, the net force will decrease. The oil will not rise up because only air is pumped out of the container. Also, the density does not change significantly for any difference to be noticed. Thus, only option A is correct.

67. If the efficiency of a 1-watt electric bulb is 10%, how many photons does it emit in one second? The wavelength of light emitted by it is 500nm.
($h = 6.625 \times 10^{-34}$)

a) 1.53×10^{17} photons
b) 2.53×10^{17} photons
c) 3.53×10^{17} photons
d) 4.53×10^{17} photons
e) 5.53×10^{17} photons

Answer: (B)

Explanation: As the bulb is 1W, if its efficiency was 100%, it would emit 1 J radiant energy in 1s. But here the efficiency is 10%, hence it emits 10^{-1} J energy in the form of light in 1s, and remaining in the form of heat.

Therefore, radiant energy obtained from the bulb is:

$1s = 10^{-1}$ J

If it consists of *n* photons then:

E = n h f (where f is frequency)

E = nhc/w (where E = energy, w = wavelength and c = speed of light)

n = E w/h c

n = $10^{-1} \times 500 \times 10^{-9} / 6.625 \times 10^{-34} \times 3 \times 10^{8}$

n = 2.53×10^{17} photons.

68. **The capacitance of a parallel plate capacitor, having 20 cm x 20 cm square plates separated by a distance of 1 mm, is:**
    a) 300 pF
    b) 350 pF
    c) 400 pF
    d) 450 pF
    e) 500 pF

    Answer: (B)

    Explanation:

    $C = \varepsilon_0 A / d$

    $C = (8.85 \times 10^{-12} \text{ F/m}) \times (400 \times 10^{-4} \text{ m}^2)/(1 \times 10^{-3} \text{ m})$

    $C = 350 \times 10^{-10}$ F

    $C = 350$ pF

69. **The speed of sound in air is 324 m/s approximately. The speed of sound in a non-porous solid:**
    a) Is greater than in air
    b) Is less than in air
    c) Can't be determined
    d) Is almost the same as in air

    Answer: (A)

    Explanation: The closely packed particles in a non-porous solid provide a medium in which disturbance or wave energies are transferred in a much faster, more efficient way. Thus, the speed of sound increases. Only option A states that.

70. The path of motion of a charged particle when a magnetic field is applied perpendicular to the velocity is:
   a) Helix
   b) Square
   c) Linear
   d) Circular
   e) Parabolic

Answer: (D)

Explanation: The magnetic field has no component along the velocity, and has its entire field directed perpendicular to it. As a result, there will be no change in the magnitude of velocity, but there will be a constant change in the direction. Thus the particle will follow a circular path, option D. Since there is no other component along any direction, the other trajectories are incorrect in this scenario.

Free Response Questions

1. A small block of mass 'm' slides on a prism of mass 'M', which in turn slides backward on the horizontal surface.
   a) Find the acceleration of the smaller block with respect to the prism.
   b) Find the acceleration of the smaller block with respect to the ground.
   c) Find the acceleration of the Prism with respect to the ground.
   d) Find the normal force acting between the two blocks.

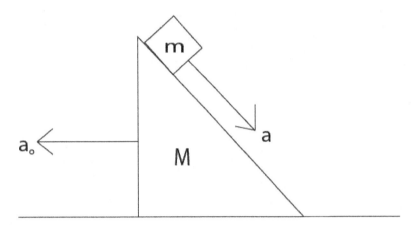

Answer:

Let the acceleration of the prism be a_0 in the backward direction and consider the motion of smaller block from the frame of the prism.

The total forces acting on the block are:

I. N (normal force)
II. mg downwards (gravity)
III. ma_0 forward (pseudo force)

The block slides down the plane, with the components of force parallel to inclination:

$$ma_0 \cos\theta + mg \sin\theta = ma$$

$$a = a_0 \cos\theta + g \sin\theta \ \ \dots \ (1)$$

(Continued on next page)

Components on the force perpendicular to incline gives:

$$N + ma_0 \sin\theta = mg \cos\theta \quad \ldots (2)$$

Now consider the motion of the prism form the ground frame. Since the frame is inertial, we do not need to consider pseudo forces. The forces acting on the prism are:

I. Mg (downward)
II. N normal to incline (due to block of mass m)
III. N' upward (due to the horizontal surface)

Equating the horizontal components, we get:

$$N \sin\theta = Ma_0$$

$$N = Ma_0 / \sin\theta \quad \ldots (3)$$

Putting in equation (2):

$$Ma_0 / \sin\theta + ma_0 \sin\theta = mg \cos\theta$$

Or:

$$a_0 = (mg \sin\theta \cdot \cos\theta)/(M + m \sin^2\theta)$$

From (1):

$$a = (mg \sin\theta \cdot \cos^2\theta)/(M + m \sin^2\theta) + g \sin\theta$$

$$a = (M g \sin\theta + mg \sin\theta)/(M + m \sin^2\theta)$$

The acceleration (A) of block 'm' with respect to ground frame is vector addition of 'a' and 'a_0'. The angle between the two vectors are (180°- θ)

$$A = \sqrt{(a^2 + a_0^2 + 2 \cdot a \cdot a_0 \cdot \cos[180° - \theta])}$$

Thus,

a) $a = (M g \sin\theta + mg \sin\theta)/(M + m \sin^2\theta)$
b) $A = \sqrt{(a^2 + a_0^2 + 2 \cdot a \cdot a_0 \cdot \cos(180° - \theta))}$
c) $a_0 = (mg \sin\theta \cdot \cos\theta)/(M + m \sin^2\theta)$
d) $N = Ma_0 / \sin\theta$

2. Two capacitors of capacitance 12 μF and 6 μF are connected in series, and the resultant is made parallel to a capacitor of capacitance 2 μF. Whole circuit is provided with a potential difference of 24 V.

   a) Find capacitance between points A and B.
   b) Find charge on the three capacitors.
   c) Taking the potential at point B to be zero, find the potential at point D.

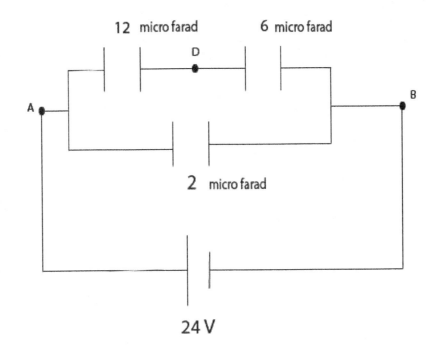

Answer:

a) The 12 μF and 6 μF capacitors are joined in series. The equivalent of these two will have a capacitance given by:

$$1/C_{eq} = 1/C_1 + 1/C_2$$

Or:

$$C = 4 \text{ μF}$$

(Continued on next page)

The combination of these two capacitors is connected in parallel with the 2 µF capacitor. Thus, equivalent capacitance between A and B is:

$$4\ \mu F + 2\ \mu F = 6\ \mu F$$

($C_{eq} = C_1 + C_2$, for parallel connection)

$$C = 6\ \mu F$$

b) The charge supplied by the battery is:

$$Q = CV$$

$$Q = 6\ \mu F \times 24\ V$$

$$144\ \mu C$$

The potential difference across 2 µF capacitor is 24 V, so, the charge on this capacitor is:

$$Q = CV$$

$$= 2\ \mu F \times 24\ V$$

$$= 48\ \mu C$$

The charge on 12 µF and 6 µF capacitor is:

$$144\ \mu - 48\ \mu$$

$$96\ \mu F$$

c) The potential differences across the 6 µF capacitor is:

$$96\ \mu F / 6\ \mu F$$

$$16\ V$$

As the potential at point B is taken as zero, the potential at point D is 16 V.

3. A player throws a basketball with an initial velocity of 20 m/s at an angle of 45° with the horizontal.
    a) How long does it take to reach the ground?
    b) What is the maximum height it reaches?
    c) How far from the player is it when it reaches the ground?
    (Take the value of g = 10 m/s², and ignore the height of player)

Answer:
a) Take the origin as the point where the ball is thrown, vertically upward as Y-axis, and horizontal in plane of motion as X-axis. The initial velocity has components:

$$U_x = (20 \text{ m/s}) \cos 45° = 10\sqrt{2} \text{ m/s}$$

$$U_y = (20 \text{ m/s}) \sin 45° = 10\sqrt{2} \text{ m/s}$$

When the ball reaches the ground, y is equal to zero.

$$Y = U_y t - 1/2\, gt^2$$

$$0 = (10\sqrt{2} \text{ m/s})t - 1/2 \times (10)t^2$$

$$t = 2\sqrt{2} \text{ s}$$

$$t = 2.8 \text{ s}$$

Thus, it takes 2.8 s for the basketball to reach the ground.

b) At its highest point, $V_y = 0$

$$V_y^2 = U_y^2 + 2 \cdot g \cdot y$$

$$0 = (10\sqrt{2} \text{ m/s})^2 + 2 \times 10 \times y$$

$$y = 10 \text{ m}$$

Thus, the maximum height reached is 10 m.

(Continued on next page)

The horizontal distance travelled before falling is given by:

$$X = U_x t + 1/2\, at^2$$

Since acceleration along X-axis is 0, thus,

$$X = U_x t$$

$$X = (10\sqrt{2}\text{ m/s})(2\sqrt{2}\text{ s})$$

$$X = 40\text{ m}$$

4. A block of mass 1kg initially at rest and kept on a horizontal smooth surface, is given a constant horizontal acceleration of 5m/s².
   a) Find the speed after first 10 seconds.
   b) Find the displacement covered in the first 10 seconds.
   c) Find the change in kinetic energy in the first 10 seconds.

Answer:
   a) v = u + at

   u = 0 m/s and a = 5 m/s² and t = 10s

   v = 50 m/s

   b) s = ut + 1/2(at²)

   s = 250m

   c) Change in K.E. is equal to the difference in initial and final Kinetic energies. Initially at rest, K.E. is zero. Finally, a speed of 50m/s, hence K.E. = 125 J

   The difference = 125 J

5. Between two parallel charged plates, a uniform electric field E is created, as illustrated in figure. Now, an electron enters the field symmetrically between the two plates with the speed V_0. The length of each of the plates is L.

    a) Find time taken in crossing the field.
    b) Find vertical component of velocity of body.
    c) Find horizontal component of velocity of body.
    d) Find the angle of deviation of the path of the electron as it comes out of the field.

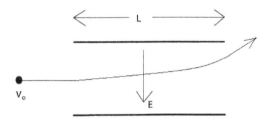

Answer:

a) The acceleration of the electron is a = eE/m in the upward direction. The horizontal velocity remains v_0, as there is no acceleration in this direction. T is the time taken to cross the field. Thus:

$$t = L/V_0$$

b) The upward component of the velocity of the electron when it emerges from the field region is:

$$v_y = at = eEL/\, m\, V_0$$

c) The horizontal component of velocity is $v_x = V_0$

d) The angle θ made by the resultant velocity with original direction is given by:

$$\tan\theta = v_y/v_x = eEL/\, mV_0^2$$

Thus, the electron deviated by an angle is:

$$\theta = \tan^{-1}(eEL/\, mV_0^2)$$

6. The half-life of ^{198}Au is 2.7 days. Calculate
    a) The decay constant
    b) The average life
    c) The activity of 1 mg of ^{198}Au.

    (Take atomic weight of ^{198}Au to be 198 g/mol)

    Answer:
    a) The half-life and the decay constant are related, as:

    $$t_{1/2} = \text{Ln } 2/\lambda = 0.693/\lambda$$

    $$\lambda = 0.693/t_{1/2} = 0.693/2.7 \text{ days}$$

    $$\lambda = 0.693/(2.7 \times 24 \times 3600 \text{ s}) = 2.9 \times 10^{-6} \text{ s}$$

    b) The average life is:

    $$t_{av} = 1/\lambda$$

    3.9 days

    c) The activity is A = λN. Now 198 g of Au has 6 ×10^{23} atoms. The number of atoms in 1 mg of Au is:

    $$N = 6 \times 10^{23} \times 1 \text{ mg}/198 \text{ g} = 3.03 \times 10^{18}$$

    Thus,

    $$A = \lambda N$$

    $$A = (2.9 \times 10^{-6} \text{ s}) \times (3.03 \times 10^{18})$$

    $$A = 8.8 \times 10^{12} \text{ disintegration/s}$$

    $$A = 8.8 \times 10^{12} / 3.7 \times 10^{10}$$

    $$A = 240 \text{ Ci.}$$

7. The displacement of a particle of a string carrying a travelling wave is
   y= (3 cm) sin 6.28(0.50x – 50t) where x is in centimeters and t is in seconds.
   a) Find the amplitude.
   b) Find the wavelength.
   c) Find the frequency.
   d) Find the speed of the wave.

   Answer:
   a) Comparing with standard wave equation:

   $$Y = A \sin(kx - \omega t)$$

   $$= A \sin(x/\lambda - t/T)$$

   Amplitude (A) = 3 cm

   b) Wavelength (λ) = 1/0.50 = 2 cm
   c) Frequency = v = 1/T = 50Hz
   d) Speed = v λ = 50 Hz x 2cm = 100 cm/s

8. The angle of incidence when light falls on a transparent material is 45° and the angle of refraction is 30°. If the incidence side medium is air, then
   a) Find the refractive index of the material.
   b) The speed of light in the material.

   Answer:
   a) The refractive index is equal to the sine of angle of incidence divided by the sine of angle of refraction. Hence:

   RI= sin i/sin r

   RI= sin 45°/sin 30° = 1.414

   b) The speed of light in that medium will be equal to the speed of light in air (the incidence side medium) divided by the refractive index. Hence, speed is equal to:

   $$c/RI = (3 \times 10^8 \, m/s)/1.414 = 2.12 \times 10^8 \, m/s$$

Quiz 3

Multiple Choice

1. **An astronaut standing on the moon drops a feather and a hammer from rest. The object that reaches the ground first is:**
   a) The hammer
   b) The feather
   c) Both of them
   d) Both float in the air
   e) Can't be determined

   Answer: (C)

   Explanation: Both the objects reach the ground at the same time, because there is no atmosphere on the moon to provide drag or resistance to the fall of the feather, and the acceleration due to the moon's gravity is the same for both the objects. Hence option C is correct, and the other options are invalid.

2. The two ends of a light spring are displaced along its length. All displacements having equal magnitude, the case in which the magnitude of tension or compression in the spring will be maximum is:

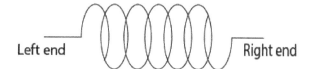

a) Left end shifts left, right end shifts right
b) Both ends shift right
c) Both ends shift left
d) Right end shifts right, left end remains stationary
e) Right end stationary, only left end shifts

Answer: (A)

Explanation: When the right end shifts right and the left end shift left, there is a net elongation, creating tension. In options B and C, the movement of both the ends is such that there is no net elongation or compression (since both the ends move in the same direction with the same magnitude of displacement). In option D and E, there is displacement only of one end, which results in lesser elongation than in case A. Hence, only option A is valid.

3. The acceleration a of the elevator such that the box of mass M exerts a force of Mg/4 on the elevator floor is:

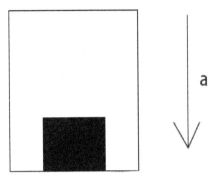

a) 4g
b) 3g/4
c) 6g
d) 1g/2
e) g

Answer: (B)

Explanation: The box is at rest with respect to the elevator which is in turn accelerated with respect to the ground. Hence, the acceleration of the block with respect to the ground is a. The forces acting on the block are Mg downwards and Normal (N) upwards. The net acceleration due to these forces is downward. We have the equation:

$$Mg - N = Ma$$

If normal is supposed to be Mg/4, the equation gives the value of a as 3g/4.

4. A long empty trolley of mass M is moving on a smooth horizontal road with a constant velocity of v. After some time, a helicopter comes and begins to drop stones, each of mass m, into the trolley at regular intervals. What will eventually happen?

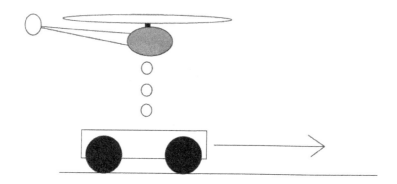

a) The speed of the trolley will be more than v
b) The speed of the trolley will be less than v
c) The speed of the trolley remains constant
d) It depends on the speed of the helicopter
e) Insufficient data

Answer: (B)

Explanation: As the helicopter drops the stones vertically, there is no force acting on the stones and trolley system in the horizontal direction. Hence, momentum is conserved in the horizontal direction. The constant value is Mv, but since the mass of the trolley is continuously increasing, its velocity will have to come down so as to nullify the effect of increased mass, so that the net momentum is still constant. Thus, after some time, the mass of the trolley will have increased, but its speed will have decreased to keep the momentum constant. Thus only option B is correct, whereas other options are incorrect.

5. **The necessary condition for an object to be in complete equilibrium (both translational and rotational) is:**
   a) Net force should be zero
   b) Net torque should be zero
   c) Net force should be zero, net torque can be non-zero
   d) Net torque should be zero, net force can be non-zero
   e) Both net force and net torque should be zero

   Answer: (E)

   Explanation: For a body to be in translational equilibrium (acceleration to be zero), the net force should be zero. And for rotational equilibrium, net torque should be zero, so that there is no angular acceleration acting on the body. In order to attain both the types of equilibrium simultaneously, the net force and the net torque acting on the body should amount to zero. Only option E is correct.

6. **When Albert Einstein shook hands with Charlie Chaplin, the significant kind of force that their hands exerted on each other is:**
   a) Electromagnetic
   b) Gravitational
   c) Weak
   d) Nuclear
   e) None of the above

   Answer: (A)

   Explanation: When they shook hands, they made contact. The contact forces that come up are primarily electromagnetic in nature, arising due to the interaction of molecules and atoms at the periphery of both the objects. This is the same kind of force responsible for normal force between two objects in contact. Thus, only option A is significant. The force is neither nuclear nor weak, which are intra-molecular and intra-atomic in nature. Gravitational force does exist, but the question demands the significant type, which is electromagnetic contact force.

7. **A block slides down an incline of angle 30° with an acceleration g/4. The coefficient of kinetic friction is:**
   a) 2/√3
   b) 1/
   c) 1/√3
   d) 1/2
   e) 1/2√3

   Answer: (E)

   Explanation: Assume the mass of the block to be M. The forces on it are:

   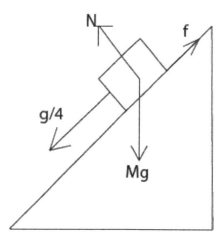

   Mg downwards, N (normal) perpendicular to the incline and f (friction) along the incline. Taking components parallel to the incline we get:

   $Mg \sin 30° - f = Mg/4$

   $f = Mg/4$

   Also, there is no acceleration perpendicular to the incline, hence:

   $N = Mg \cos 30° = Mg \cdot (\sqrt{3}/2)$

   As the block is slipping:

   $f = \mu_k N$

   Thus,

   $\mu_k = f/N = 1/2\sqrt{3}$

8. For the scenario described below, the wall is smooth but the surface of A and B in contact are rough. When the system is in equilibrium, the friction on the smaller block due to the larger block is:

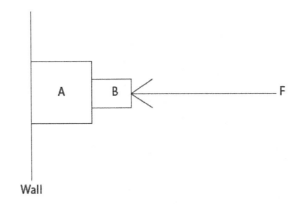

a) In the downward direction
b) In the upward direction
c) Zero
d) Equilibrium is not possible in this case
e) Insufficient data

Answer: (D)

Explanation: Since the system is vertical and the wall is smooth, there is no friction between the wall and the two-block system to hold it against gravity; as a result, regardless of the magnitude of F and friction between the blocks, the system will slide freely. The absence of a vertical force that could balance the gravitational pull is the reason that the system cannot achieve equilibrium. Hence, only option D is the correct choice.

9. A simple pendulum is constructed by attaching a bob of mass m in a string of length L. The bob oscillates in a vertical circle. It is found that the speed of the bob is v when the string makes an angle θ with the vertical. The tension in the string at this instant is:

a) m(gsinθ + v/L)
b) m(gcosθ + v²/L)
c) m(gcotθ + v²)
d) m(gtanθ + v/L)
e) m(gcotθ + v)

Answer: (B)

Explanation:

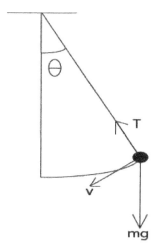

The forces acting on the bob are Tension T and weight mg. As the bob moves in a vertical circle with its center at O, taking components along the string and applying Newton's second law, we have:

$$T - mg\cos\theta = mv^2/L$$

$$T = m(g\cos\theta + v^2/L)$$

10. When a particle moves in a circle with uniform speed:
   a) Its velocity and acceleration are both constant
   b) Its velocity is constant, but the acceleration changes
   c) Its acceleration is constant, but its velocity changes
   d) Both acceleration and velocity change
   e) Neither acceleration nor velocity changes

Answer: (C)

Explanation: When the body moves in a circular path, with uniform speed, it implies that the acceleration is constant and is directed perpendicular to the direction of velocity at any instant. As a result, the direction of velocity changes but its magnitude remains constant. This is why option C is correct and the rest all are incorrect.

11. A block of mass *m* slides down a frictionless mountain of lateral height *L* and angle of incline ϑ. What is the speed of the block as it comes down to the foot of the mountain?

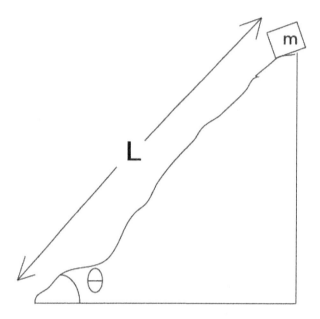

a) $\sqrt{(gL\sin\theta)}$
b) $\sqrt{(2gL\sin\theta)}$
c) $\sqrt{(3gL\sin\theta)}$
d) $\sqrt{(4gL\sin\theta)}$
e) $\sqrt{(5gL\sin\theta)}$

Answer: (B)

Explanation: Since the mountain is frictionless and the only force acting on the block is conservative in nature (gravitational force), the mechanical energy of the block + earth system is conserved. Thus, as the block descends, its potential energy gets converted to kinetic energy.

Let the speed at the bottom be *v*. The height of the block initially is $L\sin\vartheta$. Then we have:

$$mgh = \tfrac{1}{2}mv^2$$

$$mg(L\sin\theta) = \tfrac{1}{2}mv^2$$

$$v = \sqrt{(2gL\sin\theta)}$$

12. Four balls, numbered 1, 2, 3 and 4, all with different masses (as shown in the figure) are placed at the four corners of a square of side length *a*. The center of mass of this system is:

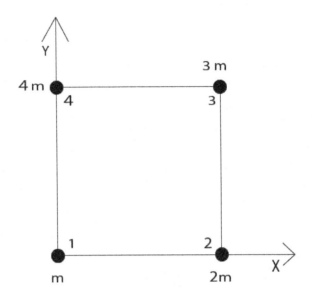

a) (a/3, a/3)
b) (a/2, a)
c) (a, a)
d) (a/5, a/6)
e) (a/2, 7a/10)

Answer: (E)

Explanation: Using the formula for locating the center of mass, we have:

$X_{com} = (m.0 + 2ma + 3ma + 4m.0)/(m + 2m + 3m + 4m) = a/2$

$Y_{com} = (m.0 + 2m.0 + 3ma + 4ma)/(m + 2m + 3m + 4m) = 7a/10$

COM is at (a/2, 7a/10).

13. **A cubical block of ice with mass *m* and edge *L* is placed in a large tray of mass *M*. If the ice melts, the distance by which the center of mass of the cube + tray system comes down is:**

a) L
b) L/2
c) mL/2(m+M)
d) 2L
e) mL/(m+M)

Answer: (C)

Explanation: When the ice melts, the water of mass *m* spreads on the surface of the large tray.

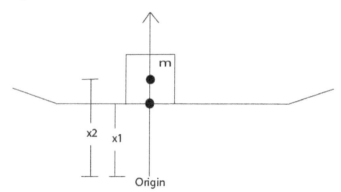

As the tray is large, the height of water is negligible. The center of mass is then on the surface of the tray and is at a distance $(x_2 - L/2)$ above the origin. The new center of mass of the ice-tray system will be at the height x', where:

$$x' = \{m(x_2 - L/2) + Mx_1\}/(m + M)$$

The shift is:

$$= x - x' = mL/2(m + M)$$

14. In a two particle system of masses m_1 and m_2, the first particle is pushed towards the center of mass by a distance d. The distance by which the second particle should be moved so as to keep the COM at the same position is:

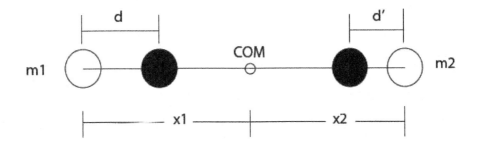

a) d
b) $d(m_1/m_2)$
c) $d(m_2/m_1)$
d) $d/2$
e) $d/4$

Answer: (B)

Explanation: Assuming the distance of m_1 from COM is x_1 and that of m_2 is x_2, and distance moved by m_2 to keep the COM at the same position d', we have:

$$m_1 x_1 = m_2 x_2$$

And:

$$m_1(x_1 - d) = m_2(x_2 - d')$$

Subtracting the second from first equation:

$$m_1 d = m_2 d'$$

$$d' = (m_1/m_2)d$$

15. An action movie star enacts a scene in which he has to jump on a moving car from above. If the mass of the car is *M* and the actor's mass is *m*, and if the speed of the car before the jump is *v*, then the speed after the jump is:
   a) v
   b) v/2
   c) (m/M)v
   d) mv/(m + M)
   e) Mv/(m + M)

Answer: (E)

Explanation: Consider the car plus the actor as the system. Since there is no horizontal external force on this system, its momentum in the horizontal direction will be conserved. Initial horizontal momentum is Mv. Let the velocity after jump be v'. Thus, the after-jump momentum will be:

$$(M + m)v'.$$

We have:

$$Mv = (M + m)v'$$

$$v' = Mv/(M + m)$$

16. In the following scenario, the person who is more likely to fall off from his seat is:

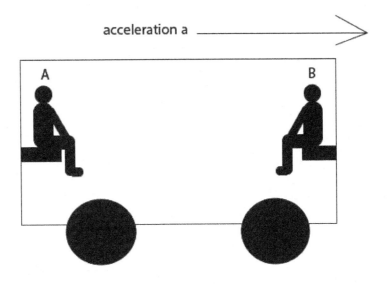

a) Person A
b) Person B
c) Both
d) None
e) Can't be determined

Answer: (B)

Explanation: Since the vehicle is accelerating towards the right, the people sitting inside also need to have the same acceleration so as to remain in their seats. For person A, the wall behind her will provide enough normal force so as to provide the desired acceleration and keep her in her seat. But for person B, he has no wall to provide this force, only friction between him and his seat. Friction between two surfaces has a limit and cannot cross a particular magnitude, so in the case of a quick acceleration, person B won't be able to remain in his seat as the friction won't be sufficient to keep him on it. Thus, only option B is correct.

17. The motor of an engine is rotating about its axis with an angular velocity of 100rev/minute. It comes to rest in 15s, after being switched off. If the angular deceleration is constant, the number of revolutions made by it before it comes to rest is:
    a) 9 rev
    b) 10.5 rev
    c) 11.5 rev
    d) 12.5 rev
    e) 13.5 rev

Answer: (D)

Explanation: The initial angular velocity:

$$100 \text{rev/min} = (10\pi/3) \text{ rad/s}$$

Final angular velocity = 0 rad/s. Assuming angular acceleration to be a, we use the following equation in rotational form:

$$v = u + at$$

Thus, we obtain:

$$a = (-2\pi/9) \text{ rad/s}^2$$

The angle rotated by the motor can be calculated using the following equation in rotational form:

$$s = ut + \tfrac{1}{2}at^2$$

Thus, we obtain rotations

$$= 25\pi \text{ rad} = 12.5 \text{ rev.}$$

18. When a ceiling fan is switched on, it takes 5 seconds for it to attain the maximum speed of 400 rpm. If the acceleration is constant, the time taken by the fan to achieve half of its maximum speed is:
    a) 0.5 sec
    b) 1.5 sec
    c) 2.5 sec
    d) 3.5 sec
    e) 4.5 sec

    Answer: (C)

    Explanation: Let the angular velocity be a. According to the question:

    $$400 \text{ rev/min} = 0 + a(5s) \quad \{v=u+at\}$$

    Taking t as the time taken to reach the speed of 200 rev/min, then:

    $$200 \text{ rev/min} = 0 + at$$

    Dividing the first equation by the second, we get:

    $$2 = (5/t) \text{ sec}$$

    $$t = 2.5 \text{ sec.}$$

19. Two uniform identical rods, each of mass M and length L are joined to form a cross section. The moment of inertia of the cross, about a bisector (dotted lines) is:

a) $ML^2/12$
b) $ML^2/6$
c) ML^2
d) $ML^2/4$
e) $ML^2/16$

Answer: (A)

Explanation: Consider an axis perpendicular to the plane of the cross and passing through its center. The moment of inertia of each rod about this line is $ML^2/12$, so the total inertia of the cross about this line is $ML^2/6$.

The moment of inertia of the cross about the two bisectors is equal by symmetry and, according to the theorem of perpendicular axes, the moment of inertia of the cross about the bisector is $ML^2/12$.

20. Two asteroids of mass 1 kg and 2 kg are floating around in outer space. If the separation between them is of 50cm and the only force acting between them is mutual gravitational pull, then the initial acceleration of the 2 kg particle is:
    a) 2.65×10^{-10} m/s²
    b) 5.35×10^{-10} m/s²
    c) 2.45×10^{-10} m/s²
    d) 5.23×10^{-10} m/s²
    e) 6.34×10^{-10} m/s²

    Answer: (A)

    Explanation: The force of gravitation exerted by one particle on the other is:

    $F = Gm_1m_2/r^2$

    $= (6.67 \times 10^{-11} \times 1\text{kg} \times 2\text{kg})/(0.5\text{m})^2$

    $= 5.3 \times 10^{-10}$ N

    The acceleration of 2 kg particle is: F/m

    $= (5.3 \times 10^{-10} \text{ N})/(2 \text{ kg})$

    $= 2.65 \times 10^{-10}$ m/s²

21. If a particle of mass 50 g, placed at a point, experiences a gravitational force of 2 N, then the gravitational field at that point is:
    a) 10 N/kg
    b) 20 N/kg
    c) 30 N/kg
    d) 40 N/kg
    e) 50 N/kg

    Answer: (D)

    Explanation: The gravitational field at any point has a magnitude = F/m.

    $E = F/m = 2 \text{ N}/50 \text{ g} = 40 \text{ N/kg}$

    This field is along the direction of the force.

22. Which of the following is false about a black hole?
   a) Even light cannot escape a black hole
   b) The escape velocity for a black hole is greater than the speed of light
   c) They are extremely dense
   d) They shine by giving out an intense amount of light
   e) They have extremely strong gravitational pull

Answer: (D)

Explanation: Black holes are extremely massive, dense bodies, whose gravitational pull is so intense that even photons of light cannot escape it. Because of this, they have an extremely high escape velocity that is even greater than the speed of light. No material particle can escape their gravity field. Since light cannot escape a black hole, they appear to be completely dark, giving out no light at all. Thus, option D is false.

23. For a projectile, the speed at the point of maximum height:
   a) Is greater than the initial speed of projection
   b) Is the point of maximum speed for the entire trajectory
   c) Is the point of minimum speed for the entire trajectory
   d) Is between the maximum and minimum values of speed
   e) Can't be determined

Answer: (C)

Explanation: At the point of maximum height, only the horizontal component of velocity remains; as a result, the magnitude of net velocity is lowest at that point and it can be concluded that the speed is minimum at the highest point for the entire trajectory. This is vouched for only by option C. The rest are incorrect.

24. If a block of mass *m* is pressed against a spring with a spring constant of *k* and compression *x*, what is the speed of the block after release? Assume the block to be moving on a smooth horizontal surface post-release.

    a) $\sqrt{(Kx)}/m$
    b) k^2x^2m
    c) $k\sqrt{m}/x^2$
    d) $x\sqrt{(k^2/m)}$
    e) $x\sqrt{(k/m)}$

Answer: (E)

Explanation: Since there is no external force in the horizontal direction, the total mechanical energy remains conserved for the system. Initially, there is only elastic potential energy, ½kx² and finally there is only kinetic energy of the block ½mv² (assuming speed to be v). But these two quantities are equal as energy is conserved throughout, thus:

½kx² = ½mv²

v = x√(k/m).

25. David ties a stone of mass *m* to a light thread of length *L* and swirls it around in a vertical circle like a sling. If at all points, the speed of the stone is *v*, then the tension in the string when the stone is at the lowest point of the vertical circle is:

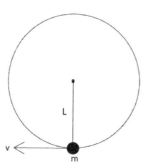

a) mg + mv²/2L
b) mg + mv²/L
c) m + mv²/3L
d) mg + mv
e) mg + mv/L

Answer: (B)

Explanation:

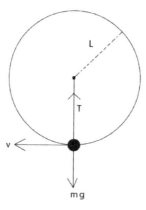

As is evident from the figure, at the lowest point, the forces acting on the stone are tension upwards and *mg* downwards. Since it is a vertical circle with a uniform speed *v*, the acceleration at this point is v²/L upwards towards the center of the circle. Thus the equation for vertical forces is:

T − mg = mv²/L

Thus, T = mg + mv²/L

26. A cylindrical vessel containing a liquid is closed by a smooth piston of mass m. The area of cross-section of the piston is A. If the atmospheric pressure is P_o, the pressure of the liquid just below the piston is:

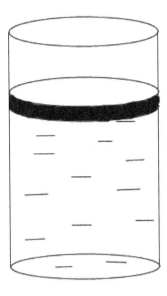

a) $P_o A$
b) $P_o + mg/A$
c) $P_o A + mg/A$
d) $P_o + mgA$
e) P_o

Answer: (B)

Explanation: Let the pressure of the liquid just below the piston be P. The forces acting on the piston are *mg* downwards, force due to atmosphere downwards ($P_o A$), and force due to liquid upwards (PA). When the piston is in equilibrium,

$PA = P_o A + mg$

$P = P_o + mg/A$

27. A liquid is placed in a cylindrical vessel, which is kept inside an elevator. The elevator has an upward acceleration a_o as shown in the figure. Thus:

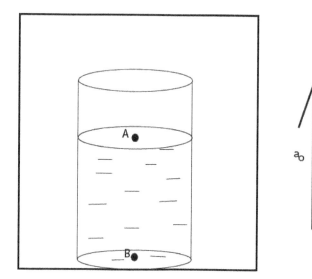

a) Point A has greater pressure than point B
b) Point B has greater pressure than point A
c) Both the points have the same pressure
d) Both the points are in the same horizontal plane
e) Insufficient data

Answer: (B)

Explanation: Since the liquid is placed inside the elevator which is accelerating at the rate of a_o, the liquid will also have an equal acceleration. The contents of the liquid just above the base of the container are accelerated due to the normal from the base and they also are forced down by the weight of the liquid above them. Whereas, at point B, near the brim of the container, the force is due to the liquid below and the atmosphere above. Clearly, the force at point B is greater than at point A and hence the pressure at point B is greater. This is explained only by option B.

28. A liquid is being pushed out of a tube by pressing a piston. The area of cross-section of the piston is 1 cm² and that of the outlet of the tube is 20 mm². If the piston is pushed at a rate of 2 cm/s, then the speed of the outgoing liquid is:

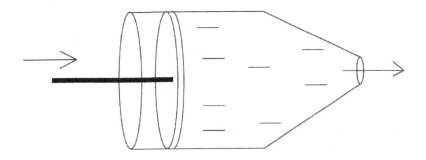

a) 6 cm/s
b) 7 cm/s
c) 8 cm/s
d) 9 cm/s
e) 10 cm/s

Answer: (E)

Explanation: From the equation of continuity, we have:

$$A_1 v_1 = A_2 v_2$$

Which is:

$$(1 \text{ cm}^2)(2 \text{ cm/s}) = (20 \text{ mm}^2)(v)$$

On solving for v, we get:

$$v = 10 \text{ cm/s}.$$

29. Three vessels A, B and C, are kept on a horizontal surface as shown. Each of them has the same base area. If an equal volume of liquid is poured into all the three, then the force on the base is maximum for:

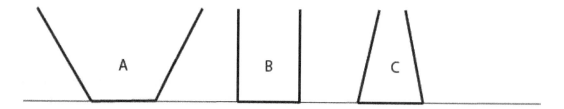

a) Vessel A
b) Vessel B
c) Vessel C
d) All have the same
e) Insufficient data

Answer: (C)

Explanation: The three vessels have the same base area but varying cross-sectional areas. In vessel C, the cross sectional area decreases with an increase in height. Thus, in order to achieve equal volume compared to the other two vessels, the height reached by the liquid will be more than the other two vessels. As a result, the pressure of the liquid at the base (which depends on the height of the liquid above the point) will be the maximum for vessel C.

When this force is multiplied by the base area, the force obtained will also be the maximum in vessel C. The height of the liquid will be in the ascending order for vessel A, B and C respectively. Same applies to the force. Hence, option C is the suitable choice.

30. Bernoulli's Theorem is based on the principle of:
    a) Conservation of mass
    b) Conservation of momentum
    c) Conservation of energy
    d) Conservation of angular momentum
    e) None of the above

    Answer: (C)

    Explanation: Bernoulli's Theorem (or principle) involves the conservation of energy. It accounts for potential difference and kinetic energy of fluid particles. It does not apply to mass or momentum.

31. A string of linear mass density 1 g/cm has a block of mass 1 kg attached to it over a light pulley. A wave pulse travels on it. The time taken by the wave pulse to travel a distance of 50 cm is:

    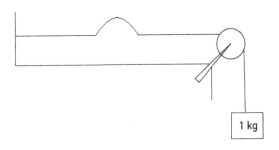

    a) 0.05s
    b) 0.10s
    c) 0.15s
    d) 0.20s
    e) 0.25s

    Answer: (A)

    Explanation: The tension in the string is T=mg=10N.

    The mass per unit length is µ=1 g/cm=0.1kg/m.

    The wave velocity is v=√(Tµ)= √(10N/0.1kg/m)=10m/s.

    The time taken to travel 50cm = (50 cm)/(10m/s) = 0.05sec.

32. The average power transmitted through a given point on a string supporting a sine wave is 0.20 W when the amplitude of the wave is 2 mm. The power transmitted through this point, if the amplitude is increased to 3 mm, is:
   a) 0.50 W
   b) 0.45 W
   c) 0.60 W
   d) 0.55 W
   e) 0.40 W

   Answer: (B)

   Explanation: If the other quantities remain constant, the power transmitted in a wave is proportional to the square of its amplitude.

   Thus:

   $P_1/P_2 = A_1^2/A_2^2$

   $P_2/0.20W = 9/4 = 2.25$

   $P_2 = 2.25 \times 0.20 = 0.45$ W

33. Concerning a sine wave travelling in a medium, what is the minimum distance between two particles always having the same speed?

   (A) $\lambda/4$
   (B) $\lambda/2$
   (C) λ
   (D) $\lambda/3$
   (E) None of the above.

   Answer: (B)

   Explanation: The key here is <u>same speed</u> and not <u>same velocity</u>. The quantity under question is a magnitude of velocity, without direction factor. At a distance of $\lambda/2$, wave particles will always have equal magnitude of velocity but different directions. Whereas at a separation of λ, they have both equal magnitude and direction. Since the question asked about the minimum distance of separation only option B is correct.

34. **Velocity of sound in air is 332 m/s. Its velocity in vacuum will be:**
    a) >332 m/s
    b) <332 m/s
    c) =332 m/s
    d) Insufficient data
    e) Meaningless

    Answer: (E)

    Explanation: Sound is a mechanical wave, so it requires a medium to travel – and vacuum is the absence of media. The question is meaningless.

35. **Which of the following require a medium to travel?**
    a) Visible light
    b) Microwaves
    c) Radio waves
    d) Radiations
    e) None of the above

    Answer: (E)

    Explanation: All of the above are non-mechanical in nature, which means that they do not need a medium.

36. **The wavelength of light coming from a sodium source is 589 nm. The wavelength in water is: (Refractive index = 1.33)**
    a) 400 nm
    b) 413 nm
    c) 423 nm
    d) 433 nm
    e) 443 nm

    Answer: (E)

    Explanation: The wavelength in water is $\lambda = \lambda_o/\mu$, where λ_o is the wavelength in vacuum and μ is the refractive index of water.

    Thus:

    $$\lambda = 589/1.33 = 443 \text{ nm}.$$

37. **When light enters a denser medium from a rarer medium:**
    a) The refracted ray bends towards the normal to the surface separating the two dissimilar optical media
    b) The refracted ray bends away from the normal to the surface separating the two dissimilar optical media
    c) The refracted ray carries on its original path with no change
    d) The refracted ray gets reflected off the surface separating the two dissimilar optical media
    e) None of the above

    Answer: (A)

    Explanation: Refractive index µ=sin i/sin r where *i* is the angle of incidence and *r* is the angle of refraction. Thus, when light enters a denser medium from a rarer medium, the refractive index is above one, which means sin *i* is greater than sin *r*. This, in turn, means that the angle of refraction is smaller than the angle of incidence, so the light ray bends towards the normal, to the surface separating the two dissimilar optical media. Only option A stands.

38. **The minimum thickness of a film which will strongly reflect the light of wavelength 589 nm is: (Refractive index of the film is 1.25)**
    a) 112 nm
    b) 114 nm
    c) 116 nm
    d) 118 nm
    e) 120 nm

    Answer: (D)

    Explanation: For strong reflection, the least optical path difference introduced by the film should be λ/2. The optical path difference between the waves reflected from the two surfaces of the film is 2µd. Thus, for strong reflection:

    $$2\mu d = \lambda/2$$

    $$d = \lambda/4\mu = 589 / (4 \times 1.25) = 118 \text{ nm}$$

39. The speed of light in a medium depends on:
   a) The elasticity of the medium only
   b) On inertia of the medium only
   c) On elasticity as well as inertia
   d) Neither on elasticity nor on inertia
   e) Latent heat of fusion

Answer: (D)

Explanation: Being a non-mechanical wave, the speed of light does not depend on the mechanical properties of the medium, namely elasticity and inertia. However, it does depend on optical characteristics like refractive index. Hence, only option D is correct.

40. The equation of a light wave is written as y = A sin(kx – ωt). Here, y represents:
   a) Displacement of ether particles
   b) Pressure in the medium
   c) Density of the medium
   d) Electric field
   e) None of the above

Answer: (D)

Explanation: Light waves are basically electromagnetic waves. Thus, an equation representing a light wave basically shows the wave motion of electromagnetic waves. It is not a mechanical wave propagating by pressure variations, particle displacement or density variations. Only option D is correct.

41. According to the laws of reflection:
   a) Angle of incidence is equal to angle of refraction
   b) Angle of incidence is equal to angle of reflection
   c) The incident ray, the reflected ray and the normal to the reflecting surface at the point of incidence all lie in the same plane
   d) Only A and B
   e) Only B and C

Answer: (D)

Explanation: The laws of reflection are:

- Angle of incidence is equal to angle of reflection
- Incident ray, the reflected ray and the normal to the reflecting surface at the point of incidence, all lie in the same plane

Only option D represents both.

42. **A convex mirror has its radius of curvature as 20 cm. The position of the image of an object placed at a distance of 12 cm from the mirror is:**
   a) (60/11) cm
   b) (62/11) cm
   c) (45/12) cm
   d) (47/12) cm
   e) (50/11) cm

   Answer: (A)

   Explanation: Since it's a convex mirror, we have:

   $$u = -12 \text{ cm}, R = +20 \text{ cm}$$

   Now,

   $$1/u + 1/v = 2/R$$

   $$1/v = 2/R - 1/u$$

   $$= (2/20 \text{cm}) - (1/-12 \text{cm}) = 11/60 \text{cm}$$

   $$v = 60/11 \text{ cm}.$$

   The positive sign shows that the image is formed behind the convex mirror, thus it is virtual in nature.

43. A printed page is kept pressed by a glass cube (μ=1.5) of edge 6 cm. The distance by which the printed words appear to be shifted, when viewed from the top, is:
   a) 1 cm
   b) 2 cm
   c) 3 cm
   d) 4 cm
   e) 5 cm

Answer: (B)

Explanation: The thickness of the cube is t=6 cm. The shift in the position of the printed letters is:

$$\Delta t = (1 - 1/\mu)t$$

$$= (1 - 1/1.5) \times 6 \text{ cm} = 2 \text{ cm}.$$

44. Which of the following expressions is known as the lens maker's formula?
   a) $1/f = (\mu - 1)(1/R_1 - 1/R_2)$
   b) $1/v - 1/u = 1/f$
   c) $1/v + 1/u = 1/f$
   d) $v + u = f$
   e) None of the above

Answer: (A)

Explanation: The expression in option B is the lens formula. The expression in option C is the mirror formula. The expression in option C is random and is not always true, and option D is too simple. The correct option is A; it is used by lens makers to design lenses with a particular focal length.

45. A biconvex lens has radii of curvature 20 cm each. If the refractive index of the material of the lens is 1.5, then the focal length is:
    a) 10 cm
    b) 15 cm
    c) 20 cm
    d) 25 cm
    e) 30 cm

    Answer: (C)

    Explanation: In a biconvex lens, the center of curvature of the first surface is on the positive side of the lens and that of the second surface is on the negative side. Thus:

    $R_1 = 20$ cm and $R_2 = -20$ cm.

    We have:

    $$1/f = (\mu-1)(1/R_1 - 1/R_2)$$

    Putting the values and solving for f, we get f = 20cm

46. The gas pressure inside a constant volume gas thermometer at steam point (373.15k) is 1.50 x 10⁴ Pa. The pressure at the triple point of water will be:
    a) 1.10 x 10⁴ Pa
    b) 2.10 x 10⁴ Pa
    c) 3.10 x 10⁴ Pa
    d) 4.10 x 10⁴ Pa
    e) 1.60 x 10⁴ Pa

    Answer: (A)

    Explanation: The temperature in kelvin is defined as:

    $$T = P/P_{tr} \times 273.16 \text{ K}$$

    Thus:

    $$373.15 = (1.50 \times 10^4 \text{ Pa} \times 273.16 \text{ K})/P_{tr}$$

    $$P_{tr} = 1.50 \times 10^4 \text{ Pa} \times 273.16/373.15$$

    $$= 1.10 \times 10^4 \text{ Pa}$$

47. If the vapor pressure of air at 20° C is found to be 12mm of Hg, what is the relative humidity? (The saturation vapor pressure (SVP) of water at 20° C is 17.5mm)
    a) 49%
    b) 59%
    c) 69%
    d) 79%
    e) 89%

    Answer: (C)

    Explanation: The saturation vapor pressure of water at 20°C is 17.5 mm of Hg. The relative humidity is equal to the vapor pressure of air divided by the SVP at the same temperature: 12 mm of Hg/17.5 mm of Hg. Thus, the relative humidity is 0.69.

48. A vessel of volume 2 liters contains 0.1 mole of oxygen and 0.2 mole of carbon dioxide. If the temperature of the mixture is 300 K, its pressure is:
   a) 2.50×10^3 Pa
   b) 1.50×10^5 Pa
   c) 4.50×10^4 Pa
   d) 5.75×10^5 Pa
   e) 3.75×10^5 Pa

Answer: (E)

Explanation: We have:

$$p = nRT/V$$

The pressure due to oxygen is:

$$P1 = (0.1 \text{ mol})(8.3 \text{ J/mol-k})(300 \text{ K})/(2000 \times 10^{-6} \text{ m}^3) = 1.25 \times 10^5 \text{ Pa}$$

Similarly, the pressure due to carbon dioxide is:

$$P2 = 2.50 \times 10^5 \text{ Pa}$$

Thus, the net pressure is:

$$(1.25 + 2.50) \times 10^5 \text{ Pa} = 3.75 \times 10^5 \text{ Pa}$$

49. When a metal sheet with a circular hole is heated, the hole:
   a) Gets larger
   b) Gets smaller
   c) Stays the same
   d) Gets deformed
   e) Insufficient data

Answer: (A)

Explanation: Since the metal expands upon heating, the various linear dimensions also show a change. During heating, metal expands, so the radius of the inner hole also increases. Thus, only option A is correct.

50. A gas is contained in a vessel fitted with a movable piston. The container is placed on a hot stove. A total of 100 calories of heat is given to the gas, and the gas does 40 J of work in the expansion resulting from heating. The increase in internal energy in the process is:

a) 350 J
b) 358 J
c) 360 J
d) 378 J
e) 388 J

Answer: (C)

Explanation: Heat given to the gas is:

$\Delta Q = 100 \text{cal} = 418 \text{ J}$

Work done by the gas is:

$\Delta W = 40 \text{ J}$

The increase in internal energy is:

$\Delta U = \Delta Q - \Delta W = 418 \text{ J} - 40 \text{ J} = 378 \text{ J}$

51. A process is said to be adiabatic when:
   a) Pressure remains constant
   b) Volume remains constant
   c) Temperature remains constant
   d) No heat is exchanged with the surroundings
   e) None of the above

Answer: (D)

Explanation: An adiabatic process is one in which heat exchange with the surrounding is allowed. The heat content remains the same. The options A, B and C describe isobaric, isochoric and isothermal processes respectively.

52. The process of heat transfer through radiation:
   a) Needs a material medium
   b) Needs no material medium
   c) Does not occur at all
   d) Is opposite of nuclear fusion
   e) None of the above

Answer: (B)

Explanation: The process of heat transfer through radiation needs no material medium as heat in this case travels in the form of electromagnetic radiations, which are non-mechanical waves, requiring no medium for their propagation. Hence, only option B is correct.

53. The absorptive power of a body is:
   a) Not a valid quantity
   b) A dimensionless quantity
   c) Defined as the energy incident divided by energy absorbed
   d) Is a unit of energy
   e) None of the above

Answer: (B)

Explanation: Absorptive power is defined as the ratio of absorbed power to the incident power. Since it is a ratio, it is a dimensionless quantity. So only option B is correct; option C is just the opposite.

54. The light from the sun is found to have a maximum intensity near the wavelength of 470 nm. Assuming that the surface of the sun emits as a blackbody, calculate the temperature of the surface of the sun.
    a) 6100k
    b) 6110k
    c) 6120k
    d) 6130k
    e) 6140k

    Answer: (D)

    Explanation: For a blackbody:

    $\lambda_m T = 0.288$ cm-k

    Thus,

    $T = (0.288 \text{cm-k})/470\text{nm} = 6130$ K

55. The material conductivity of a rod depends on its:
    a) Length
    b) Area of cross-section
    c) Mass
    d) Material of the rod
    e) None of the above

    Answer: (D)

    Explanation: Thermal conductivity is a constant for various materials at a particular temperature. It depends on the mechanical properties of the material rather than its dimensions like mass, length or area. Hence, only option D is correct.

56. **The smallest magnitude of charge(negative or positive) that can exist is:**
    a) 1.602×10^{-19} C
    b) 1.502×10^{-19} C
    c) 1.402×10^{-19} C
    d) 1.302×10^{-19} C
    e) 1.202×10^{-19} C

    Answer: (A)

    Explanation: Charge can exist only in integral multiples of *e* whose magnitude is 1.602×10^{-19} C. This is known as quantization of electric charge. Hence only option A is correct.

57. **In electrostatics:**
    a) There can be an electric field inside a conductor
    b) There cannot be an electric field inside a conductor
    c) Conductors don't exist
    d) Insulators don't exist
    e) None of the above

    Answer: (B)

    Explanation: In electrostatics, whenever a conductor is placed inside an electric field, the electrons inside the conductor rearrange themselves in such a way such that they create a secondary electric field inside the conductor which is equal in magnitude but opposite in direction to the external electric field. As a result, there is no net electric field inside the conductor. Thus only option B is correct.

58. **When a positive charge is taken from a low potential region to a high potential region, the electric potential energy:**
    a) Increases
    b) Decreases
    c) Remains constant
    d) May increase or decrease
    e) None of the above

    Answer: (A)

    Explanation: Whenever a positive charge is taken from a region of lower potential to a region of higher potential, positive work is done. This positive work gets stored in the system and hence its potential increases. Thus, only option A is correct.

59. **If a body is charged by rubbing, its weight:**
    a) Remains precisely constant
    b) Decreases slightly
    c) Increases slightly
    d) May increase slightly or decrease slightly depending on charge
    e) None of the above

    Answer: (D)

    Explanation: The body may either lose electrons while rubbing and become positive or it may gain electrons and become negative. Depending upon the situation, it either undergoes a slight decrement in weight or slight increment, respectively. Hence it depends on the charge. Thus, only option D is correct.

60. A metallic particle having no net charge is placed near a metallic plate having a positive charge. The electric force on the particle will be:
   a) Zero
   b) Parallel to the plate
   c) Away from the plate
   d) Towards the plate
   e) None of the above

   Answer: (D)

   Explanation: The electric force between a charged particle and a neutral particle is always attractive in nature. Thus, the metallic particle gets attracted towards the plate, meaning the electric force on it acts towards the metallic plate.

61. Two particles A and B of charges 8 x10⁻⁶ C and -2 x10⁻⁶ C respectively, are kept at a separation of 20 cm. The position where the third charge C should be placed, so that it does not experience any kind of electric force is:

a) 20 cm from B, 40 cm from A on line AB
b) 30 cm from B, 50 cm from A on line AB.
c) C, lies in between A and B, 10 cm from each A and B.
d) 20 cm from A, 40 cm from B on line AB.
e) 30 cm from A, 50 cm from B on line AB.

Answer: (A)

Explanation: As the net electric force on C should be equal to zero, the force due to A and B must be equal and opposite in direction. So, the particle should be placed on the line AB. As A and B have charges of opposite sign, C cannot be between A and B. A has larger magnitude of charge than B, so C should be placed closer to B than A. So let BC be X and the charge on C be Q.

Force due to A = $(8 \times 10^{-6} \text{ C})Q/(4\pi\epsilon_0(20+X)^2)$

Force due to B = $(20 \times 10^{-6} \text{ C})Q/(4\pi\epsilon_0 X^2)$

They are oppositely directed and have to be zero, so their magnitude must be equal:

$8/(20+X)^2 = 2/X^2$

On solving the quadratic equation in X, we will get:

$X = 20$ cm.

62. In a particular circuit, 10 C of charge is passed through a battery in a given time. The plates of battery are maintained at a constant potential of 12 V. The work done by the battery is:

a) 100 J
b) 120 J
c) 140 J
d) 160 J
e) 180 J

Answer: (B)

Explanation: By definition, the work done to transfer a charge q through a potential difference of V is Vq. Thus, the work done by the battery is:

$$Vq = 10 \text{ C} \times 12 \text{ V} = 120 \text{ J}$$

63. Two charges +10 μC and +20 μC are placed at a distance of 2 cm. The electric potential due to the pair at the middle point of the line joining the two charges is:

a) 17 V
b) 27 V
c) 37 V
d) 47 V
e) 57 V

Answer: (B)

Explanation: Using the equation:

$$V = Q/(4\pi\varepsilon_0 r)$$

The potential due to +10 μC is:

$$V_i = (10 \times 10^{-6} \text{ C}) \times (9 \times 10^9 \text{ N-m}^2/\text{C}^2) / 10^2 \text{ m} = 9 \text{ MV}$$

The potential due to 20 μC is:

$$V_i = (20 \times 10^{-6} \text{ C}) \times (9 \times 10^9 \text{ N-m}^2/\text{C}^2) / 10^2 \text{ m} = 18 \text{ MV}$$

The net potential at the midpoint is:

$$9 \text{ MV} + 18 \text{ MV} = 27 \text{ MV}$$

64. Two particles, each of mass 5g and charge 1 x10⁷ C, stay in limiting equilibrium on a horizontal table with a separation of 10 cm between them. The coefficient of friction between each particle and the table is the same. The value of the coefficient is:
   a) 0.16
   b) 0.17
   c) 0.18
   d) 0.19
   e) 1.00

Answer: (C)

Explanation: The electric force on one of the particles due to other is:

$$F = (9 \times 10^9 \text{ N}^2\text{-m}^2/\text{C}^2) \times (1 \times 10^{-7} \text{ C}^2) \times (0.1 \text{ m})^{-2} = 0.009 \text{ N}$$

The frictional force is in limiting equilibrium

$$f = \mu mg$$

$$= \mu \times 5 \times 10^{-3} \text{ kg} \times 9.8 \text{ m/s}^2$$

$$= \mu \times 0.049 \text{ N}$$

As these two forces balance each other,

$$0.009 = 0.049 \times \mu$$

Thus,

$$\mu = 0.18$$

65. The figure shows electric field lines corresponding to an electric field. The figure suggests that:

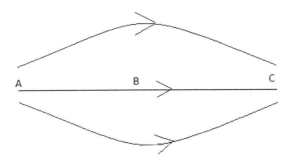

a) Ea > Eb > Ec
b) Ea > Ec = Eb
c) Ea = Eb > Ec
d) Ea = Ec > Eb
e) None of the above

Answer: (D)

Explanation: The field line density at point A and point C is the same, so electric field strength at these two points are equal. Now, the density of the electric field at point C is less than that at A and C. So the electric field strength at will be less than that of A and C. Thus,

$Ea = Ec > Eb.$

66. Two capacitors each having capacitance C and breakdown voltage V are joined in series. The total capacitance and breakdown voltage will be:

a) 2C and 2V
b) C/2 and V/2
c) 2C and V/2
d) C/2 and 2V
e) C and V

Answer: (D)

Explanation: Since the two capacitors are connected in series…

$$1/C_{eq} = 1/C_1 + 1/C_2$$

$$C_{eq} = (1/C + 1/C)^{-1}$$

$$C_{eq} = C/2$$

Now the charge given by the battery of potential will be constant in the circuit, and let it be q:

$$C = q/V$$

$$q = CV \ldots (1)$$

Let the new potential be V'. Then,

$$C/2 = q/V'$$

$$q = CV'/2 \ldots (2)$$

On equating equation 1 and 2,

$$CV'/2 = CV$$

$$V' = 2V$$

67. **A beam consisting of protons and electrons moving at the same speed passes through a thin region containing a magnetic field perpendicular to the beam. The protons and electrons:**
   a) Will go un-deviated
   b) Will be deviated by same angle but will not separate
   c) Will be deviated by different angles and will separate
   d) Will be deviated by same angle but will separate
   e) None of the above

Answer: (C)

Explanation: Since the charges are moving perpendicular to the magnetic field, forces will be applied on both the charges due to the field.

However, since he charges are unlike, the direction of force applied is not the same, and since the masses of the proton and electron are not the same, their accelerations are also different. Therefore, the deviation of both the particles is different.

68. **The radius of Li^{++} ions in its ground state assuming Bohr's model is valid, is:**
   a) 8pm
   b) 18pm
   c) 28pm
   d) 38pm
   e) 48pm

Answer: (B)

Explanation: For hydrogen-like ions, the radius of the *nth* orbit is:

$a_n = n^2 a_0 / Z$

For Li^{++}, Z = 3 and in ground state n =1, the radius is:

$a_1 = 53pm/3 = 18$ pm

69. How many different wavelengths may be observed in the spectrum from a hydrogen sample if the atoms are excited to states with principal quantum number *n*?
    a) $n^2(n-1)/2$
    b) $n^2(n+1)/2$
    c) $n(n-1)/2$
    d) $n/2$
    e) $n^2(n-1)$

    Answer: (C)

    Explanation: From the nth state, the atom may go to (n-1)th state, ..., 2nd state or 1st state. So there are (n-1) possible transitions starting from the nth state. The atoms reaching (n-1)th state may make (n-2) different transitions. Similarly for other lower states. The total number of possible transitions is:

    $$(n-1) + (n-2) + (n-3) + \ldots 2 + 1 = n(n-1)/2$$

70. A vertical electric field of magnitude 4.00×10^5 N/C just prevents a water droplet of mass 1.00×10^{-4} kg from falling. The charge on the droplet is:
    a) 1.45×10^{-9} C
    b) 2.45×10^{-9} C
    c) 3.45×10^{-9} C
    d) 4.45×10^{-9} C
    e) 5.45×10^{-9} C

    Answer: (B)

    Explanation: The forces acting on a droplet are the electric force qE vertically upwards and the force of gravity mg downwards

    To just prevent it from falling, these two forces should be equal and opposite.

    $q(4.00 \times 10^5 \text{ N/C}) = (1.00 \times 10^{-4} \text{ kg}) \times (9.8 \text{ m/s}^2)$

    $q = 2.45 \times 10^{-9}$ C

Free Response

1. A pulley of radius *r* and mass *M* is fixed at its center by a clamp. A light rope is wound over it and the free end is tied to a block. The tension in the rope is *T*.

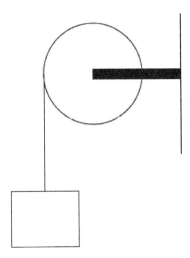

a) What are the forces on the pulley and the relation between them?
b) Where is the axis of rotation?
c) What is the torque of the forces about the axis of rotation?

Answer:

A. The forces on the pulley are: (i) attraction by the earth, Mg, vertically downwards. (ii) Tension T by the rope, along the rope. (iii) Contact force N by the support at the center. The center of mass of the pulley is at rest, hence Newton's first law applies:

$$N = T + Mg$$

B. The axis of rotation is the line through the center of the pulley and perpendicular to the plane of the pulley.

C. The force Mg passes through the center of the pulley, which means it passes through the axis, hence its torque about this axis is zero.

Same applies to the force N. No torque. The tension T is along the tangent of the rim in the vertically downward direction. The tension and the axis of rotation are perpendicular and never intersect. Hence its torque about the axis is T.r (positive).

2. The ladder shown in the figure has negligible mass and rests on a frictionless floor. The crossbar connects the two legs of the ladder at the middle. The angle between the two legs is 60°. The person sitting on the ladder has a mass of 80 kg.

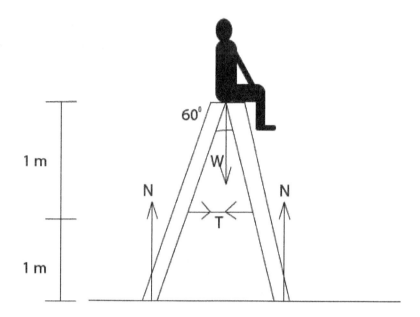

a) What is the contact exerted by the floor on each leg?
b) What is the tension in the crossbar?

Answer:

A. The forces acting on the different parts are shown in the figure itself. Considering the vertical equilibrium of "the ladder + the person" system, the forces acting are the two normal and the weight:

$2N = W$

$N = 40 \text{ kg} \times 9.8 \text{ m/s}^2 = 392 \text{ N}$

The force exerted by the floor on the legs is 392 N.

B. On considering the equilibrium of the left leg of the ladder; taking torques of the forces acting on it about the upper end:

$N (2m) \tan 30° = T(1m)$

$T = N \times 2/\sqrt{3} = 392 \times 2/\sqrt{3} = 450 \text{ N}$

3. From the graph given below, find:

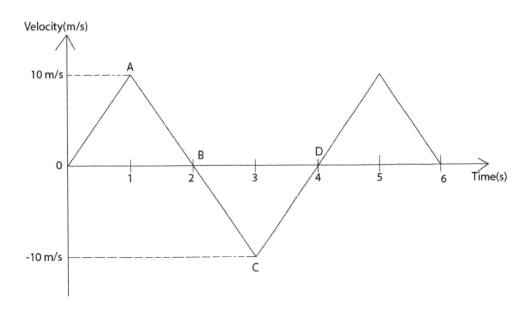

a) The distance traveled during the first two seconds
b) During 2s to 4s
c) Displacement during 0s to 4s
d) Acceleration at t=1/2 sec
e) Acceleration at t=2 sec

Answer:

A. Distance during 0 to 2 s is area of OAB = ½ x 2s x 10m/s = 10m.

B. Distance during 2 to 4 s is area of BCD = 10m. But the particle has moved in the negative x-direction.

C. Net displacement from 0s to 4s = displacement from 0s to 2s + displacement from 2s to 4s = (10m) + (-10m) = 0m.

D. At t = ½ sec, acceleration = slope of line OA = 10 m/s^2

E. At t = 2 sec, acceleration = slope of line ABC = -10 m/s^2

4. A block of mass 10 kg is kept on a rough horizontal surface with a co-efficient of static friction at 0.75. A force of 100N is applied to it on the horizontal direction.
   a) What is the magnitude of limiting friction?
   b) Show that the block moves.
   c) If the block moves what is the acceleration of the motion?

Answer:

A. The magnitude of limiting friction $f = \mu N$, where N is the normal force.

$$N = 10 \text{ kg} \times g = 100 \text{ N}$$

And so $f = 0.75 \times 100N = 75$ N. This is the maximum limit for friction, hence limiting friction.

B. Since the applied force is greater than the opposing friction, the body will not stay in a state of equilibrium and will move.

C. The net force acting on the block is equal to:

$$F - f = 100 - 75 = 25 \text{ N}.$$

$$F' = ma = 10 \text{ kg} \times a = 25 \text{ N}$$

Solving for a, we get:

$$a = 2.5 \text{ m/s}^2$$

5. The electron energy levels are as follows for an atom: K-Level = (-3 x 10⁻¹⁵ J), L-level = (-3 x 10⁻¹⁶ J) and M-level = (-3 x 10⁻¹⁷ J). The atom is bombarded with high energy electrons. The impact of one of these electrons has caused the complete removal of k-level electrons and replacement with an electron of the L-level with a certain amount of energy being released during the transition. This energy may appear as x-rays or may all be used to eject an M-level electron from the atom.

   (A) What is the minimum potential difference through which the electron may be accelerated from rest to cause the ejection of K-level electron from the atom?

   (B) What Is the energy released when L-level electron moves to fill the vacancy in K-level?

   (C) What is the wavelength of the x-ray emitted?

   (D) What is the kinetic energy of the electron emitted from the M-level?

   Answer:

   A. $eV = 3 \times 10^{-15}$ J

   $V = (3 \times 10^{-15} \text{ J})/(1.6 \times 10^{-19}) = 1.875 \times 10^4$ V

   B. Energy released is:

   $\Delta E = (3 \times 10^{-15} - 3 \times 10^{-16}) = 2.7 \times 10^{-15}$ J

   C. $\lambda = (12400 \times 1.6 \times 10^{-19})/(2.7 \times 10^{-15}) = 0.737 \times 10^{-10}$ m

   D. Released energy is:

   2.7×10^{-15} J.

   This energy also releases M electron so, KE:

   $= 2.7 \times 10^{-15} - 3 \times 10^{-17}$

   $= 10^{-15} [2.7 - 0.03] = 2.67 \times 10^{-15}$ J

6. A sample of an ideal gas is taken through a cyclic process, *abca*. It absorbs 50 J of heat during part *ab*, no heat during part *bc* and rejects 70 J of heat during part *ca*. 40 J of work is done on the gas in part bc.

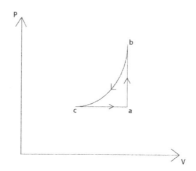

a) Find the internal energy of the gas at *b* and *c* if it is 1500 J at *a*.
b) Calculate the work done by the gas during part *ca*.

Answer:

A. In *ab*, the volume remains constant so the work done by the gas is zero. The gas absorbs 50 J of heat. The increase in internal energy from *a* to *b* is:

$$\Delta U = \Delta Q = 50 \text{ J}$$

As the internal energy is 1500 J at *a*, it will be 1550 J at *b*. In part *bc*, the work done by the gas is $\Delta W = -40$ J and no heat is given to the system. The increase in internal energy from *b* to *c* is:

$$\Delta U = -\Delta W = 40 \text{ J}$$

As the internal energy is 1550 J at *b*, it will be 1590 J at *c*.

B. The change in internal energy from *c* to *a* is:

$$\Delta U = 1500 \text{ J} - 1590 \text{ J} = -90 \text{ J}$$

The heat given to the system is $\Delta Q = -70$ J

Using the following equation:

$$\Delta Q = \Delta U + \Delta W$$

We get:

$$\Delta W = \Delta Q - \Delta U = -70 \text{ J} + 90 \text{ J} = 20 \text{ J}$$

7. A sample of an ideal gas (γ = 1.4) is heated at constant pressure. If an amount of 140 J of heat is supplied to the gas, then:
   a) Find the change in the internal energy of the gas.
   b) Find the work done by the gas.

   Answer:

   Suppose the sample contains n moles and also suppose the volume changes from V_1 to V_2 and the temperature changes from T_1 to T_2.

   A. The change in internal energy is:

   $$\Delta U = nC_v (T_2 - T_1) = C_v/C_p \times nC_p \times (T_2 - T_1)$$

   $$= C_v/C_p \times \Delta Q = 140/1.4 = 100 \text{ J}$$

   B. The work done by the gas is:

   $$\Delta W = \Delta Q - \Delta U = 140 \text{ J} - 100 \text{ J} = 40 \text{ J}$$

8. A battery of emf 2V and internal resistance 0.50 Ω supplies a current of 100 mA.
   a) What is the potential difference across the terminals of the battery?
   b) How much thermal energy develops in the battery in 10 seconds?

   Answer:

   A. The potential difference across the terminals is:

   $$\Delta V = E - ir$$

   $$= 2 \text{ V} - (0.100 \text{ A})(0.50 \text{ Ω}) = 1.95 \text{ V}$$

   B. The thermal energy developed in the battery is:

   $$U = i^2 rt = (0.100 \text{ A})^2 (0.50 \text{ Ω})(10\text{s}) = 0.05 \text{ J}$$

Image Credits

1. "Third Law," Pg. 24; modified from original Microsoft Office clip art. Used with permission from Microsoft.

2. "Projectile Motion," Pg. 35; original artwork by author; "paper airplane" object in image modified from original Microsoft Office clip art. Used with permission from Microsoft.

3. "Torque," Pg. 60; original artwork by author; "wrench" object in image modified from original Microsoft Office clip art. Used with permission from Microsoft.

4. "Electronic Symbols," Pg. 90; public domain image accessed at: http://en.wikipedia.org/wiki/Electronic_symbol#mediaviewer/File:Circuit_elements.svg

5. "Right Hand Rule and Magnetic Fields," Pg. 104; public domain image accessed at: http://commons.wikimedia.org/wiki/File:Right_hand_rule.png

6. "Sine and Cosine," Pg. 110; public domain image accessed at: http://en.wikipedia.org/wiki/Sine_wave#mediaviewer/File:Sine_and_Cosine.svg

7. "The Electromagnetic Spectrum," Pg. 116; public domain image accessed at: http://mynasadata.larc.nasa.gov/science-processes/electromagnetic-diagram/

8. "Two-Slit Experiment Particles," Pg. 117; public domain image accessed at: http://commons.wikimedia.org/wiki/File:Two-Slit_Experiment_Particles.svg

9. "Two-Slit Experiment Light," Pg. 117; public domain image accessed at: http://commons.wikimedia.org/wiki/File:Two-Slit_Experiment_Light.svg

10. "Lenses," Pg. 126; public domain image accessed at: http://www.docstoc.com/docs/127133864/Lenses---PCC

11. "Atom," Pg. 131; public domain image accessed at: http://www.wpclipart.com/science/atoms_molecules/atom_diagram.png.html

12. "Atom Orbital," Pg. 131; public domain image accessed at: http://en.wikipedia.org/wiki/Atomic_orbital#mediaviewer/File:Neon_orbitals.JPG